Maryland's Geology

Maryland's Geology

by Martin F. Schmidt, Jr.

To Leigh —

Hope you enjoy the geology wherever you travel

Martin F. Schmidt, Jr.
2/27/13

4880 Lower Valley Road, Atglen, PA 19310

Published by Schiffer Publishing, Ltd. 2010
Maryland's Geology was originally published by Tidewater Publishers in 1993
Copyright © Tidewater Publishers in 1993
Copyright Martin F. Schmidt, Jr. 2010

Library of Congress Control Number: 2010920881

First edition, 1993; third printing, 2010

ISBN: 978-0-7643-3593-8
Printed in China

Schiffer Books are available at special discounts for bulk purchases for sales promotions or premiums. Special editions, including personalized covers, corporate imprints, and excerpts can be created in large quantities for special needs. For more information contact the publisher:

Published by Schiffer Publishing Ltd.
4880 Lower Valley Road
Atglen, PA 19310
Phone: (610) 593-1777; Fax: (610) 593-2002
E-mail: Info@schifferbooks.com

For the largest selection of fine reference books on this and related subjects, please visit our web site at
www.schifferbooks.com

We are always looking for people to write books on new and related subjects. If you have an idea for a book please contact us at the above address.

This book may be purchased from the publisher.
Include $5.00 for shipping.
Please try your bookstore first.
You may write for a free catalog.

In Europe, Schiffer books are distributed by
Bushwood Books
6 Marksbury Ave.
Kew Gardens
Surrey TW9 4JF England
Phone: 44 (0) 20 8392 8585; Fax: 44 (0) 20 8392 9876
E-mail: info@bushwoodbooks.co.uk
Website: www.bushwoodbooks.co.uk

All topographic maps are adapted from maps by the Maryland or U.S. Geological Survey. All other illustrations are by the author, excepted as noted.

Cover photograph of Sideling Hill in Washington County by J. Paul Breeding. Courtesy Maryland Geological Survey.

Contents

Invitation and Introduction

We live on a fascinating and active planet. North America has a particularly large selection of interesting geological features, and Maryland, since it contains a cross section of the Appalachian Mountains, has an especially varied geology for a rather small state. So, there's a lot to learn about and understand for anyone ready to look, and that's what this book is about. It's an open invitation to take a look in some depth—into the ground and back in time—at the part of the earth we call Maryland.

This book is written so you don't have to know much about geology when you start. If an introduction to a subject is needed in order to understand a specific area of Maryland, that introduction is provided here. While the book cannot be a complete geology course by any means, it is comprehensive enough to enable you to understand the geology of Maryland, and perhaps even apply some of the general principles to other places you visit or read about as well. At the same time, if you already know the basics of geology and only want to learn some Maryland specifics, you can skip the introductory sections in each chapter.

A look at the table of contents will give you an idea of how the book is organized. The first five chapters go into increasing detail, building somewhat on each other: chapter 1 simply describes surface landforms; chapters 2 and 3 look beneath the surface at the underlying rocks and structures to find the cause of the landforms; and chapters 4 and 5 look back through time to see how the rocks and structures formed. Finally, chapter 6 examines how the geology of Maryland affects people, both in providing what we need and in sometimes limiting what we can do.

Acknowledgments

This book does not represent original research on my part; however, it does pull together many facts and interpretations that cannot be found in one place otherwise. I must acknowledge my own teachers and colleagues for what they taught me about geology and Maryland; I hope this book will teach others. Reviews provided by John Jedlicka, Gary Vom Lehn, Robert C. Smoot III, and Wallace White were all helpful, and review by Loretta L. Molitor was especially valuable. I also want to acknowledge the help and support of many people at the Maryland Geological Survey in putting this book together. Specifically, John Edwards, Jr., was particularly generous with his knowledge, time, and resources, which were most helpful. I appreciate the time and comments provided by Jim Reger and Ken Schwarz, as they helped improve both the geological contents and the clarity of presentation.

Maryland's Geology

Fig. 1-1. Satellite picture of Maryland. Source: EarthSat Corporation, Chevy Chase, MD.

Chapter One

The Landforms of Maryland

Maryland contains a rather wide variety of landscapes for a state its size. This variety is especially apparent when we look at the spectacular and beautiful satellite view shown in Figure 1-1. Ours is the first generation to see a picture of a whole state at once, and the contrasts among different areas of the state show up well. This variety in landforms exists mostly because the state cuts directly across the Appalachian Mountain range, which includes several distinct areas. With this diversity to deal with, it will help if we can classify different regions of the state in an orderly way.

All of North America can be divided into different regions based on the landforms and the geology of the varied areas. This kind of classification results in areas that are called physiographic divisions, which are further subdivided into physiographic provinces. The provinces for the eastern United States are shown in Figure 1-2. We will be examining the Appalachian Plateaus Valley and Ridge, Blue Ridge, and Piedmont provinces (which are all part of the Appalachian Highlands division), plus the Coastal Plain and Continental Shelf provinces (which make up the Atlantic Plain division). Looking at the area around Maryland, many of the boundaries between these provinces can be traced on the satellite photo in Figure 1-1. Finally, concentrating on Maryland alone, we can see where the province boundaries fall within the state in Figure 1-3. Major features of the land provinces are summarized in Appendix A, which also contains information that will be covered in other chapters of this book. Though local landforms vary, the general nature of each of the provinces is the same throughout. Therefore, many of the discussions about the landforms and geologic structures of Maryland will also apply to other states that have the same provinces.

The Coastal Plain Province

The main part of the Atlantic Plain of which most of us are aware is the dry land section, the Coastal Plain province. The area of Maryland that is east of the Chesapeake Bay represents classic Coastal Plain land—nearly flat

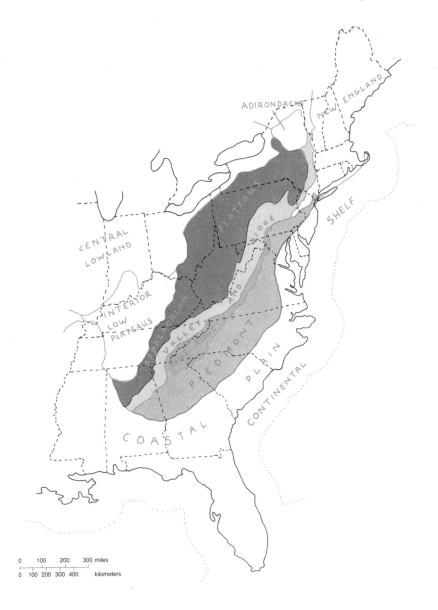

Fig. 1-2. Physiographic provinces of the eastern United States. Adapted from a map by the U.S. Geological Survey.

and nowhere much above sea level. Geologists call the difference between the highest elevation and the lowest elevation of a region the *relief* of that

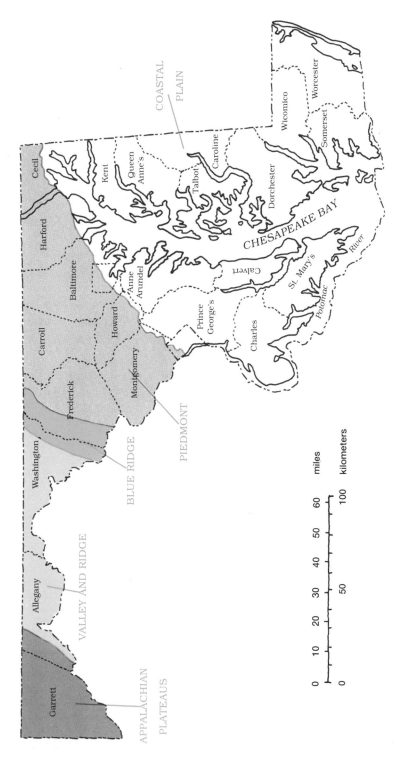

Fig. 1-3. Physiographic provinces of Maryland. Adapted from a map by the Maryland Geological Survey.

area. The Coastal Plain province is said to be of *low relief* since the high
spots aren't much above the low ones, especially east of the Chesapeake
Bay. The area of the Coastal Plain west of the Bay is slightly more hilly, and
has more relief. (A good place to see some of this higher area is at Calvert
Cliffs State Park, where the land surface clearly rises a good bit more above
sea level, that is, above Bay level, than it does, for example, at Cambridge
or Ocean City.) Still, nearly all of the Coastal Plain is of lower relief than are
the more hilly areas to the west. (Maps that show the major mountains and
hills are provided in this book for some of Maryland's provinces [Figures 1-5
through 1-8]. No map is included for the Coastal Plain because the relief is
so low that the small hills and valleys wouldn't show up well on a map that
could fit in the book. However, named places can be located by using the
geologic map of the Coastal Plain, Figure 3-1.)

The other part of the Atlantic Plain division is the Continental Shelf
province. This is the area under the ocean that starts at the coast and
stretches out into the Atlantic for about 70 miles. Like the Coastal Plain
next to it, the land surface of this underwater province has little relief. Also,
the ocean bed slopes so very gently out to sea in the Continental Shelf that,
some 70 miles out, the water depth is only about 700 feet. While that may
sound very deep to those of us used to the amount of water in a swimming
pool, it means the slope of the ocean floor is dropping only about 10 feet for
each mile it goes out to sea. To illustrate this, imagine you ran a string across
a 20-foot-wide room, and made the string drop from one end to the other at
the same rate the Continental Shelf drops. One end would be less than a
half inch lower than the other. It would be hard to tell whether such a string
was sloping at all, which shows how nearly flat the Continental Shelf is. At
the edge of the Continental Shelf, the ocean floor drops off suddenly—at a
rate of more than 400 feet for each mile—so it's some 40 times steeper than
the Continental Shelf. This steeper area is called the Continental Slope, and,
to geologists, represents the real edge of the continent. Thus, the Coastal
Plain and the Continental Shelf can be thought of as the visible and invisible
parts of the same piece of land.

Some parts of the Continental Shelf are not at all close to level, however.
Towards the seaward edge, there are a few deep canyons cut into the
surface. These canyons slope down until they reach the bottom edge of the
shelf, and there they open out to the deep ocean. While we will never see
these canyons, their existence tells us something about the geologic history
of Maryland; their origin is discussed in Chapter 5.

One other feature of the Coastal Plain lies between the land and the sea:
the coast itself. Most of the mid-Atlantic seacoast has long, narrow islands

offshore, called barrier islands or barrier beaches, each with a body of water called a lagoon between the island and the main shore. In Maryland, the barrier islands are Fenwick Island on the north, with Ocean City on it, and Assawoman and Sinepuxent bays behind it, and Assateague Island to the south, backed by Chincoteague Bay. These islands are low in elevation (less than twenty feet above sea level) and are made of sand, with bare sand dunes in some areas and low brushy vegetation covering other areas.

The Piedmont Province

All of the area west of the Coastal Plain in Maryland falls in the Appalachian Highlands division. This area contains four distinct provinces, which we'll consider individually, still moving from east to west.

West of the Coastal Plain lies the Piedmont province, whose name literally means "foot of the mountains." At such a location, we would expect the land to become hilly, and indeed it does. All of the Piedmont is generally covered by rolling hills, with one particularly high spot at Sugarloaf Mountain. The valley that contains the city of Frederick, though generally flatter and at a lower elevation than most of the province, is also part of the Piedmont. The relief of this province is too low overall to show well on a map covering the whole area, but most place names are given in Figure 3-10.

In central Maryland the increase in relief is gradual as you go west from the Coastal Plain into the Piedmont, so the boundary between the two is hard to place when looking out across the land from the ground. This boundary—called the Fall Zone—is a line connecting the areas on rivers where people traveling upstream by boat from the ocean encounter the first major rapids. In colonial times boats had to stop at this point on the river, so many towns were established in the mid-Atlantic states at or near the Fall Zone, including Trenton, Philadelphia, Wilmington, Baltimore, Washington, and Richmond. Thus, a line running through these cities roughly shows the Piedmont–Coastal Plain boundary. Interestingly, there really is a line of sorts between these cities: roads were constructed between them in colonial times and remain to this day. (The newest road to follow the Fall Zone is Interstate 95, from New Jersey through Virginia.)

To locate the boundary on the satellite photo, look at the rivers flowing into the Chesapeake Bay—each is narrow for much of its course, but then widens as it approaches the Bay. The wider section is called an *estuary*, and can be distinguished from the rest of the river by the fact that the tides rise and fall in the estuary, but not in the river. (The head of the estuary on the Potomac River, for example, is near Washington, DC.) A line connecting the places where the rivers widen into estuaries generally marks the boundary between the Piedmont and the Coastal Plain.

In addition to the hills found in both provinces, the western Coastal Plain and the Piedmont share another common landform—the pattern of the rivers and their tributaries. The pattern is visible in the satellite picture: the rivers create a branching effect like that shown in Figure 1-4A. This pattern looks like a tree, and so is called *dendritic,* from the Greek word for tree. The pattern is visible in the satellite photo, Figure 1-1, and in the maps of these two provinces, Figures 3-1 and 3-10. On the Piedmont map, the pattern is also reflected in the shapes of the reservoirs, especially Liberty Reservoir, northwest of Baltimore. On a larger scale, the whole Chesapeake Bay shows the pattern, looking like a somewhat gnarled and bent tree with stubby branches; its riverlike shape hints at its origin, which we'll come back to in Chapter 5. The dendritic pattern of the rivers in the eastern part of the state is a landform that is different from what we find farther west.

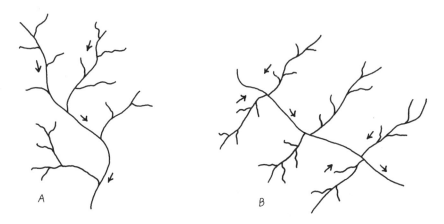

Fig. 1-4. Stream patterns: A. Dendritic pattern, B. Trellis pattern. In both, the arrows show the direction of water flow.

The Blue Ridge Province

Moving west from the Piedmont province we reach the Blue Ridge province. The change to mountainous land is obvious both when viewed from the ground, say, driving west out of Frederick, and when seen from above in the satellite picture. Part of the reason it stands out so well in the satellite photo

Fig. 1-5. Topographic map of the Blue Ridge province.

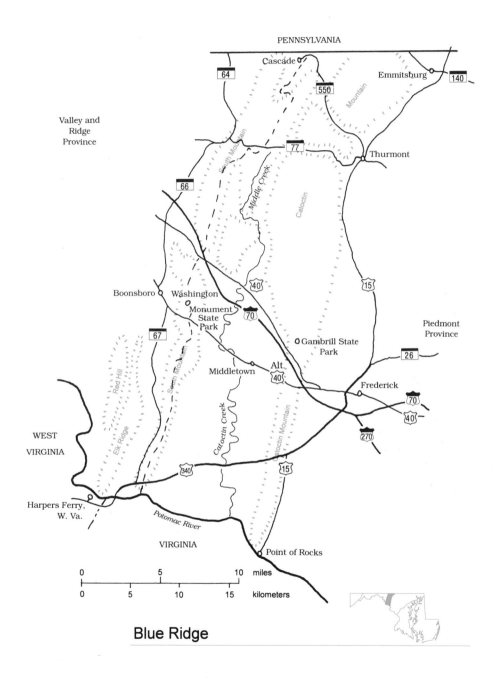

PENNSYLVANIA

Cascade

Emmitsburg

64

550

140

Valley and
Ridge
Province

South Mountain

Middle Creek

Catoctin Mountain

77

Thurmont

66

40

15

Boonsboro

Washington
Monument
State
Park

70

Gambrill State
Park

Piedmont
Province

67

Middletown

Alt.
40

26

Red Hill

South Mountain

Catoctin Creek

Catoctin Mountain

Frederick

70

40

WEST

VIRGINIA

Elk Ridge

340

15

270

Harpers Ferry,
W. Va.

Potomac River

VIRGINIA

Point of Rocks

0   5   10   miles

0   5   10   15   kilometers

Blue Ridge

is that its steep slopes are unsuitable for farming, and so remain covered with trees, which show up darker in the photo. Even without this effect, the noticeable shadows in the northern section of the Blue Ridge would indicate the area has more relief than the Piedmont.

Figure 1-5 is a topographic map of the area, that is, it shows the major landform features of the province. In most of the province, the Blue Ridge has two main ridges: South Mountain on the west, and Catoctin Mountain on the east, with the Middletown Valley between. A particularly good place to see these ridges is from the High Knob overlook in Gambrill State Park, though they are also visible from many places in the Middletown Valley. To the north, the two main ridges continue, but the area between them is also mountainous. Just west of the south end of South Mountain is a third ridge, made up of Elk Ridge and smaller Red Hill. Elk Ridge is also part of the Blue Ridge province, and it is this ridge that crosses the Potomac near Harpers Ferry, WV, and continues south as the Blue Ridge in Virginia. In both Maryland and Virginia, the Blue Ridge province is narrow, about ten miles wide, but it broadens to the south and so includes much of western North Carolina.

The Valley and Ridge Province

Continuing west from the Blue Ridge, the next province is the Valley and Ridge. The name of this province describes it well, since it consists of long linear mountain ridges, which run parallel to each other, with straight valleys between the ridges. All of the ridges and valleys run generally northeast to southwest across the state. Just west of the Blue Ridge is the widest of the valleys, known in Maryland as the Hagerstown Valley, as it contains that city. This valley also continues north and south of Maryland, and in its entirety is called the Great Valley; in Pennsylvania it's called the Cumberland Valley, and in Virginia the Shenandoah Valley. People have found the Great Valley a convenient natural highway, allowing roads such as Interstate 81 to go from Pennsylvania to Tennessee while crossing a minimum number of mountains. (A feature of the Great Valley in Maryland that is particularly noticeable in the satellite image is the meandering, or winding, path of Conococheague Creek as it flows south to the Potomac River.)

The rest of the Valley and Ridge province west of the Hagerstown Valley consists of more closely spaced ridges, with much narrower valleys. They can be seen in the satellite photo, and the mountains are named on the topographic maps, Figures 1-6 and 1-7. Overall, the province extends from

Fig. 1-6. Topographic map of the eastern Valley and Ridge province.

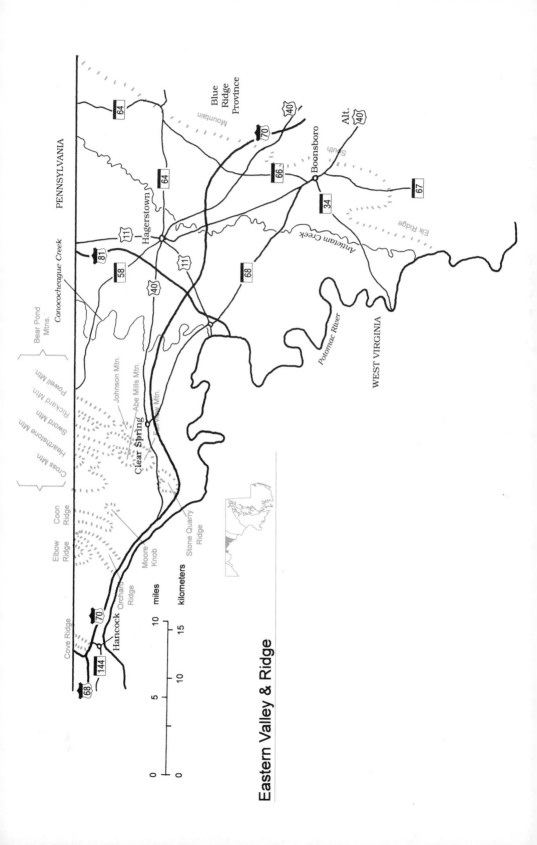

Eastern Valley & Ridge

the base of South Mountain to the base of Dans Mountain. If you stood at the top of one of the ridges of the province to view the scenery on a clear day, you would notice two things about the landscape: first, each ridge is rather even in elevation, without many high or low spots along it; second, neighboring ridges are nearly all the same elevation. Geologists call these *accordant summits,* and this term is very appropriate to the mountains of the Valley and Ridge.

The mountains here generate a different stream pattern than the one we saw earlier. The tributary rivers of the Potomac only flow in the linear valleys, not across the mountains. And they only join the Potomac in pairs, one from the north and one from the south, as the Potomac cuts west to east through each valley. This results in the pattern shown in Figure 1-4B, called a trellis pattern because it looks like the lattice of a garden trellis. We don't see the complete pattern well in Maryland alone because it includes only one side of the Potomac River. But in many places, for each stream that flows into the Potomac on the Maryland side, there is one on the other side as well. Interestingly, even a road map of the Valley and Ridge province shows this pattern, since roads run the length of most valleys, and meet at right angles the occasional road that crosses the mountain ridges.

The Appalachian Plateaus Province

The westernmost province in the Appalachian Highlands is the Appalachian Plateaus; it is divided into seven sections. All of the plateau region in Maryland is contained in the section called the Allegheny Plateau, so we will use this name for the province for the rest of the book. A map of the plateau in Maryland is shown in Figure 1-8.

A plateau generally means a wide, high area, with flat land at least in some places, not sharply pointed mountain peaks or narrow ridges. Driving west across Maryland, when we encounter the Plateau as we climb Dans Mountain just west of Cumberland, it seems like just another mountain of the Valley and Ridge. The difference is that we don't immediately drop down into another valley—we stay high because we have climbed what's known as the Allegheny Front and are now on the Plateau. From here, it continues west through Maryland and on across much of West Virginia. (Other sections of the Appalachian Plateaus province have different names, such as the Catskill, Kanawha, or Cumberland Plateau, and others, but the landforms are similar.)

Fig. 1-7. Topographic map of the western Valley and Ridge province.

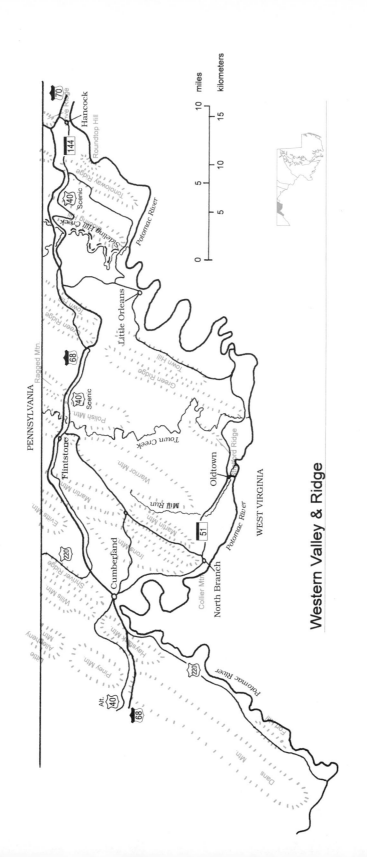

Western Valley & Ridge

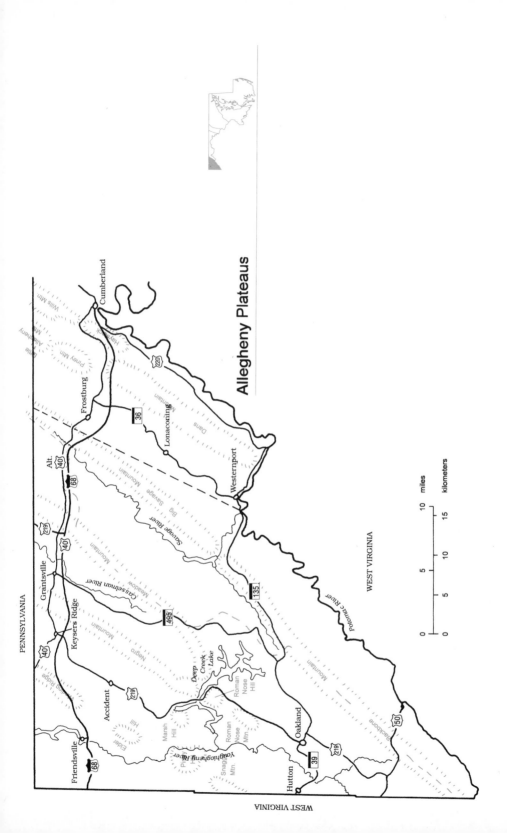

Allegheny Plateaus

The Allegheny Plateau is definitely not all flat land, however. First, there are some high ridges, with the same northeast-to-southwest trend we saw in the Valley and Ridge. As the satellite picture shows, however, the areas between ridges are noticeably wider in the Plateau. Second, the rivers and streams that cross the Plateau have cut sharply and deeply into its surface, making steep-sided valleys. Geologists call this type a *dissected plateau* because it has been so cut up by streams. In fact, it is so sliced up that most of the land is in the slope of various stream valleys, and only the accordant summits of the highlands show where the original Plateau was. We also find there is a lot of elevation change from the top of the Plateau to the bottom of the stream valleys, and so it is an area of high relief.

At the same time, the streams of the Plateau haven't been forced into parallel linear valleys as much as in the Valley and Ridge. Thus, streams in the Plateau again form the irregularly branching dendritic pattern we saw earlier. This change in stream pattern is visible on the satellite image, and helps to locate the boundary between the Valley and Ridge and the Plateau. As in the Piedmont, after people build a dam on one of these streams, the lake that fills in the valley also shows the dendritic pattern. Deep Creek Lake is a good example in the Plateau region.

Figure 1-8 includes a dashed line that is the Eastern Continental Divide. This line separates the rivers that drain to the Atlantic from those that drain to the Gulf of Mexico. Thus, the water that falls as rain and snow in northwestern Garrett County has a long trip to the Gulf via the Youghiogheny, Ohio, and Mississippi rivers, while water east of the divide has a relatively short trip to the Atlantic via the Potomac or other rivers. At the West Virginia state line, Backbone Mountain reaches the highest elevation in Maryland (3,360 feet), so it is not surprising that the Continental Divide runs along Backbone in that area. But farther north, the drainage area of the Savage River is east of the divide because the Savage has cut through the Backbone–Big Savage ridge, and drains to the Potomac. So, the divide swings northwest to Meadow Mountain for about ten miles, then returns to the Big Savage Mountain area.

Our overview of Maryland's landforms—from the mountains to the sea— is now complete. While a mere description of those landforms is informative, especially if you haven't visited all areas of the state, it is even more interesting to look at *why* each area is different. Why is one place flat and another mountainous? Why are the mountainous areas different from each other? These are the questions we'll answer in Chapter 2.

Fig. 1-8. Topographic map of the Allegheny Plateaus province.

Chapter Two

Geologic Processes That Shaped Maryland's Landscape

Having looked at the landforms of Maryland in Chapter 1, we are now about knee-deep in a branch of geology called geomorphology, and ready to go for a further plunge. As its name implies ("geo-" means earth and "morph-" means shape), geomorphology is the study of the shape of the surface of the earth. Geomorphologists often begin the study of a particular area, as we did in Chapter 1, with descriptions and classifications of the landforms. The collection of landforms found in a particular place is then called its landscape. But a geomorphologist goes on to study the processes that create the landforms, to see why they exist. This study is the part of the field we will pursue now, and it involves digging a bit deeper, both literally and figuratively.

To understand the causes of the specifics of Maryland landforms, we must first understand the generalities of how landforms are created anywhere. While some readers may be familiar with these general processes, others may not be, so we'll start with a little basic geology in this chapter. If you're already well versed in the operating rules of landform development, feel free to skip ahead to the Maryland specifics in Chapter 3.

Landform Development—The Big Picture

There are two major factors that determine the type of landforms that will develop at a particular place: the rate and type of *weathering*, and rock *character*. Weathering is the change in rocks that occurs when they are exposed to the weather—sun or rain, heat or cold, and even simple contact with the air. These changes nearly always break rocks down in one way or another, either into smaller pieces or into material which is no longer hard rock, such as soil. The second factor, rock character, as used here, includes all facts about the rock itself at a particular location: what type of rock it is, its precise composition, whether its layers are flat or folded, if it's broken or not, and more. Using these two factors, we can account for the developmen of the landforms of nearly any place on earth.

An analogy to humans may help to show how these factors interact to create landforms. Rock character is similar to heredity (the looks and other characteristics we inherit from our parents), and weathering is similar to environmental effects on people (the way everything around us affects us as we live). Rock character, like heredity, determines what the starting material is, and puts some limits on what the final product can be. For example, the softest rock in a particular area probably isn't going to form a mountain, just as parents who are five feet tall probably won't have children who grow to six feet—it's just not the right starting material for that outcome. Meanwhile, the climate and therefore the amount and type of weathering, like the environment on people, shapes the land so that even two different rocks can make similar landforms. For example, all the mountains in Maryland are rounded, rather than sharp like those in a dry climate such as Arizona. Though different Maryland mountains contain different rocks, their shapes are similar because similar weathering goes on throughout Maryland. In the same way, a brother and sister in a family, though unique individuals, also commonly share some traits because they grew up in the same home environment. In these cases in which two factors work toward the final product, both are operating at the same time, but it's easier to separate them in order to study them. That's what we'll do in the next two sections of the chapter.

Weathering: The Story of Strong and Weak Rocks

Minerals

To understand how rocks change due to the effects of the weather, we need to know what materials are in them—and this will help us out with the character of rocks also. Rocks are made of minerals, which are naturally occurring solid elements or compounds (combinations of elements) in the earth. There are thousands of types of minerals, but, fortunately for our study, only a few of them make up most of the volume of rock at the earth's surface. We call these few minerals *common minerals,* and looking at how they weather will help us to understand a lot about landforms.

A large number of minerals contain the elements silicon and oxygen in varying amounts, and are called silicates. One way of handling the common silicates is to divide them into felsic minerals and mafic minerals, as shown in Table 1. Felsic minerals are silicates that also contain elements such as aluminum, potassium, or sodium, among others; most of these minerals are light in color. Mafic minerals are silicates that also contain (especially) magnesium and iron—they're easy to remember because the name mafic comes from the start of the word magnesium, plus the first letter of the chemical symbol for iron, Fe. Most mafic minerals are dark in color, and more dense than felsic minerals. These two types of silicate minerals are widespread in Maryland and throughout the crust (the outer layer) of the earth.

TABLE 1: COMMON SILICATE MINERALS	
FELSIC MINERALS	MAFIC MINERALS
quartz	olivine
feldspars:	pyroxene
orthoclase (contain potassium)	amphibole
plagioclase (contain sodium, calcium)	biotite (dark mica)
muscovite (clear or light-colored mica)	

Another class of mineral common at the surface of the earth is the carbonates. Carbonates contain carbon and oxygen, and at least one other element. Most commonly, the other element is calcium, making the mineral called calcite. Occasionally, both calcium and magnesium are present, making the mineral dolomite. There are many other carbonates, but these are the main ones.

Rocks are made up of minerals, and in order to understand how rocks weather, it's important to know not only *what* minerals are in them, but also *how* the minerals are put together. The different ways rocks are made are reflected in the three major types: igneous, sedimentary, and metamorphic. These types, and the kinds of rocks found in Maryland within each type, are shown and briefly summarized in Appendix B. If this information (and there's a lot of it!) is new to you, you may want to read only the description of the three major rock types for now, and look up individual rocks as you encounter them in the text.

With that bit of background on the materials involved, we can now look at how these materials are affected by the weathering processes. There are many processes that occur, but they can be grouped into two general types: mechanical weathering and chemical weathering. These two kinds of weathering often occur at the same time, but it will be easier if we first look at them separately.

Mechanical Weathering

Mechanical weathering (which is also called physical weathering) is simply the breaking down of a rock or mineral into smaller pieces of the same material. One common way this occurs is through what is known as *frost action.* This happens when water enters an existing crack in a rock, then freezes and so expands, cracking the rock farther apart. Then water can move deeper into the rock the next time it thaws, and the process continues. It may seem surprising that ice could break a rock, but large amounts of pressure can build up in such cracks. We also see the results of this process when roads or driveways lift and crack during the winter.

Frost action weathers rock throughout Maryland because all areas of the state get freezing weather part of the year. It probably occurs to an added degree on the tops of mountains and in the Allegheny Plateau province of western Maryland, because the climate there is colder and there are more freeze-and-thaw cycles in a year. Frost action results in sharp-edged (a shape called *angular*) chunks of rock. A field of such blocks can be found on the tops of some mountains. An easily accessible example in Maryland is at Washington Monument State Park, though that block field is also partially the result of the construction of the monument itself.

Another form of weathering is called exfoliation, a word which comes from two Latin roots, "ex-" meaning out of or from, and "folia" meaning leaves. Exfoliation is the removal or breaking up of a rock into leaves or layers which did not exist before. This process occurs in rocks that have no layering to begin with (such rocks are said to be *massive*), such as granite or some sandstones.

Here's an example of mechanical weathering by exfoliation: Deep in the earth, as granite forms, it would be put under great pressure from above, and so be compressed. When the rocks above are removed by erosion and the pressure is relieved, the granite expands enough to crack into layers parallel to the surface of the land. The cracked pieces can then fall apart more easily than if the rock were still massive. There are a few granites in Maryland that show this type of cracking (for example, the Woodstock Granite).

One other way that rocks get broken apart is through the action of living things. Trees and other plants can pry rocks apart as their roots grow in cracks in the rocks. Burrowing animals push on cracked rocks and remove broken rock pieces. An important result of this and the other types of mechanical weathering is to increase the surface area on which chemical weathering can begin to act.

Chemical Weathering

Chemical weathering is the change in a rock due to a chemical reaction between the rock (really, the minerals in the rock) and something in the environment, such as water, oxygen, or carbon dioxide. This chemical reaction results in one or more new materials, and the original rock no longer exists afterward—it has a new form. This new form may be a solid, such as clay or iron oxide, or it may be a compound that dissolves in rainwater and so is washed away. Some specific examples with minerals will probably make this clear.

One basic chemical weathering reaction is called *hydrolysis*, in which there is a reaction between water and the minerals in a rock. This change involves not simply the dissolving, but rather the breaking-down of some of the water molecules, which then recombine with some elements of the original minerals to make new minerals. For example, this process changes feldspars and micas into clay, plus a number of elements in solution to be carried away by runoff water. When you pick up a rock and find you can crumble it in your hand, hydrolysis has probably been at work turning the hard minerals into softer, soil-like ones. Hydrolysis is one of the most common types of chemical weathering, forming the abundant clay which is a major part of many soils.

Another chemical weathering method is called *carbonation*, as it is a reaction with carbon, usually combined with oxygen as carbon dioxide. Carbon dioxide exists in the atmosphere, and is also given off by bacteria in the soil, so it is easy for rainwater to become mixed with it in the air or in the ground. When it does, it forms a weak acid called carbonic acid; this reaction explains why even unpolluted rainwater or well water is commonly slightly acidic. If rainwater containing carbonic acid comes in contact with the mineral calcite (of which limestone is made), it slowly dissolves the calcite, and the resulting ions are carried away in runoff water. Thus, where there was rock before, later there is nothing; even whole mountains of limestone can be gradually washed to the sea. (By the way, the dissolved material might become new limestone through chemical processes in the ocean, so the cycle of rock formation and weathering can continue.)

A third important chemical weathering method is *oxidation*, that is, a reaction between the minerals and the oxygen in the air or in water. This reaction is the process that causes steel objects, which are mostly made of iron, to rust when they're left out in the weather. As we know, rusting tends to make iron objects fall apart, so it's easy to see how this weathering can break down a rock. The same thing happens to the iron contained in the mafic minerals mentioned above (Table 1), once the iron is removed from the silicate by processes such as hydrolysis. Oxidation of the freed iron creates rusty brown and red iron oxides that are a major source of the brown and red colors we see in the soil. Interestingly, the rusty material may later be redeposited elsewhere to make new, red, sedimentary rocks. Indeed, certain rocks in Maryland are called "red beds" because they are layered shales formed from mud that contains enough oxidized iron to cause a notable red color. Other elements besides iron also can oxidize, contributing to the breakdown of rocks that contain these elements.

We can sum up chemical weathering by looking at what minerals are least susceptible to this type of weathering—they're called *resistant*—and which minerals are most susceptible—they're called *nonresistant*. In general, quartz, besides being physically hard, also chemically weathers very slowly, so quartz is a rather resistant mineral. It is what is left after other minerals have worn away, and continues to exist as chunks of quartz in the soil, and as sand, which is often mostly quartz, on beaches. The other felsic minerals tend to break down into clays, but are at least moderately resistant. Mafic minerals are somewhat less resistant than felsics, especially in Maryland, where the moderately wet climate provides ample moisture to promote chemical weathering. Also among the less resistant minerals are the carbonates—they slowly dissolve under the rainfall conditions of Maryland. Of course, we are describing here how easily these minerals weather compared to each other; by human standards of time, they all take a long time to weather and change.

Note that chemical and mechanical weathering occur together and affect each other. For example, oxidation, a chemical process, may cause iron-bearing compounds to detach from their original material, which is a mechanical process. This happens in natural materials, and in man-made ones, such as the rusting of steel. Similarly, when feldspars turn to clay, the material reacts with water so that the new material takes up more volume than the original material did. Then, grains or layers of the rock may loosen because they don't fit, mechanically breaking the rock apart. This sometimes results in rocks weathering apart in layers, which is another form of exfoliation. This occurs in some nonlayered rocks, with some good examples in the Piedmont. In general, while it is convenient to talk about mechanical and chemical weathering separately, they really are closely related.

Results of Weathering

All of these weathering processes result in rock falling apart, somewhat like a dead tree as it rots and disappears. This may be a new concept to some people: we normally think of stone as being solid and unchanging. But the truth is that rock, like everything else, changes and disintegrates when conditions cause it to do so. These changes have gone on worldwide and in Maryland for billions of years, and continue to go on today. It should also be mentioned that not only do our varied landforms result from this process, but also our very lives depend on it. If weathering didn't break the hard rock into soil and elements dissolved in water, there would be no materials free to grow plants and to form animal bodies. Humans are made partly of materials derived from rocks, and if it weren't for weathering, those materials would still just be hard rock. We are truly earthlings, because we're made of elements from the earth.

How mechanical and chemical weathering interact depends on some of the qualities of individual rock types. Many of the rock descriptions in Appendix B mention whether a rock is fine- or coarse-grained, and whether or not it's in layers; these qualities (and others) geologists call the *texture* of the rock. Texture affects weathering in various ways. For example, the large, visible mineral crystals in a coarse-grained rock have proportionally less surface area compared to their volume than the small crystals in a fine-grained rock. Since chemical weathering occurs especially on the surface of a grain, the weathering goes on faster in small-grained rocks, because more surface is exposed. Another texture effect is that of layering: rocks that are formed in layers may split easily into these layers again, and so weather faster than nonlayered ones.

We can bring our discussion of weathering to these conclusions: some rocks are resistant to weathering and therefore change slowly, while others are nonresistant, and they disappear more easily. This separation is important because, other things being equal, we would expect the resistant rocks to stand out as high places of the land, and the nonresistant ones, having been worn away, to be lower. But notice we can only talk about resistance by comparing one rock with another, as ultimately *all* rocks do break down. A summary of the relative resistances of rocks in Maryland is given in Appendix C. It should be noted that this summary is correct for Maryland; it would be different if it were for a different climate, such as might be found in Utah (where it's much drier), or in Brazil (where it's much warmer and wetter).

Actually, the term "weathering" that we've been using refers to what happens to the rocks *in place;* that is, the material does not move very far. Once the rock is softened into clay, or turned into material dissolved in water, it is carried off in the process called *erosion.* Erosion goes on nearly everywhere on land, in both obvious and subtle ways. The material on all slopes, no matter how gentle they are, moves downhill due to gravity and running water. Once the weathered material makes it to a stream, the stream acts as a conveyor belt to carry it away; at the same time, the stream erodes the streambed itself. These processes mean that the dominant factor shaping the land in Maryland and many other places is water, acting through both weathering and erosion. The material of the hills slowly but inevitably moves downhill and is carried away down the streams, with the whole process happening faster on nonresistant rocks and slower on resistant ones. In this way, weathering and erosion make landforms.

Now we know which rocks we'd expect to be resistant and thus make high landforms, and which wouldn't; and we know what the effects of the weather should be on the rocks in the state. But what rocks are where, and how are they arranged? To answer that, we must look at the structure of the layers of rock in Maryland.

Rock Structure: The Bends and Breaks in Rocks

Introduction

Rock structure, as the term is used in this book, refers to the shapes resulting from rocks being deformed (or not deformed) by tilting, folding, or faulting. This section of the chapter deals with kinds of rock structures in a general way, and how to find them using geologic maps. Then we will concentrate on the structures actually found in Maryland in Chapter 3.

When sediments are deposited to make sedimentary rocks, they usually accumulate in flat, horizontal layers. Even after the sediments have turned into solid rock, these layers remain and are called *strata,* or, more generally, *beds.* Whether these layers have stayed horizontal or not is an important part of the structure of an area. If they're not horizontal, they may have been moved to make a fold, or were broken and then moved, which is called faulting. (Such folding and faulting also can be seen in the layering of metamorphic rocks. Igneous rocks show other structures, some of which are mentioned below.) In any case, it is the folding, or lack of it, in sedimentary rocks that we'll mainly be concerned with here.

Geologic Maps—At the same time we discuss the possible structures, it will help if we discuss a tool for recognizing and locating the structures, and that tool is a geologic map. Geologic maps show the bedrock of an area, that is, the types of rocks you would find if all the soil and loose material above the solid rock were removed. Different types of rocks are shown by different colors or symbols, so geologic maps may be rather colorful and beautiful to look at. (The nice part is that the attractive patterns also mean something and give us some information, as long as we can read the message in the pattern.)

One pattern on the map is easy to understand, and that's the legend, or key. The legend explains what color or pattern represents which rock type. To be specific, geologic maps separate the rocks into units called *formations*, which may be one particular type of rock or a group of rocks which are convenient to map together, and which are assigned different colors on the map. These formations are also given names, which usually have two parts. The first part is a place name, usually the place or area where the rock is found, or where it was first studied. The second part is the name of the rock type; if there's a group of types, the name may be the type that occurs most, or this part may be left out completely. Examples of the usual type are Baltimore Gneiss, or Frederick Limestone. An example in which there are several rock types is the Juniata Formation, which contains siltstone, shale, and other sedimentary rocks. Such individual rock types within a formation are called *members*, and sometimes the members are named also. In some situations, two or more formations are considered together and called a *group.*

In the legend of a geologic map, the formations are usually listed in a column with the oldest formation on the bottom, and the youngest at the top. They are listed in this order because that is the way the rocks (especially sedimentary ones) are usually found on the earth. This situation makes sense, since, in any stack of objects, normally the object on the bottom was placed there first, and is the oldest part of the stack, and others above it were placed there later and thus are younger parts. This idea applies so often that it is a basic geologic principle, called the principle of superposition, and it can be applied to most rocks anywhere (though there are exceptions to it).

Another way to tell the ages of rocks on a geologic map is the geologic time periods, which may be listed beside the formations. The amount of time from when the earth formed to the present is very long—more than 4.5 billion years. We call this *geologic time* to separate it from the much shorter *historical time,* since humans have kept written records. We'll talk more about geologic time in Chapter 4, but for now you should know that geologic time has been divided up and different parts given names. These names and the times in years that go with them are listed in Appendix D. There are four long intervals of time called eras, and each era has been divided into several periods. The names of the periods usually appear next to the formation names on geologic maps, to show the time period when the rock formed. The ages of formations on the geologic maps in this book can be found in Appendix E or F. The main point here is that sometimes it is important to know which rock on a geologic map is older and which is younger. You can determine this either by looking at where the formations are in the column of rock on the legend, or by looking at what geologic time periods they are in.

Horizontal Rock Layers

With that introduction to geologic maps, we're ready to look at rock structures and how they appear on the maps. We'll assume that the rocks we're talking about started as horizontal layers, as mentioned before. As a first case to examine, the easiest to consider is one in which the beds remain horizontal. While this is a simple situation, it actually leads to a distinctive and frequently intricate pattern on geologic maps. The pattern is formed by the streams that cut into the layers, as follows: Imagine a sloping section of land with horizontal strata, as shown in Figure 2-1A. Suppose a stream cuts through the slope, so it looks like Figure 2-1B. Notice how the horizontal bed in the center seems to go back into the hill, or in the direction which is up the stream, in order to cross the stream without changing elevation. You may notice some mountain roads follow a similar in-and-out path, so that they don't have to go up- or downhill much to cross a stream running down the side of a mountain. When viewed directly from above, which is the view used for a geologic map, the rock bed would seem to make a V shape that points upstream. The V will actually be rounded off because real land doesn't have the sharp corners that Figure 2-1B has.

That's the pattern created by crossing *one* stream, but of course a landscape like Maryland's actually contains many streams, and so might look like the bottom block of Figure 2-2. The view you would see from directly above the block is shown at the top of Figure 2-2, and this is the geologic map view. Now the rounded V's from many streams get connected to each other, and the pattern of colors looks somewhat like the outline of a leaf or a branching bush. We saw a branching pattern before in streams and tributaries, and it has the same name here: it is a *dendritic pattern.* Thus, horizontal rock layers result in a dendritic pattern, and anytime we see that

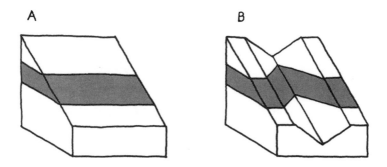

Fig. 2-1. Block diagrams showing a horizontal rock layer: A. Sloping land, B. Sloping land with a stream valley cut into it.

pattern in a geologic map we can conclude that the rocks are flat-lying. A place where bedrock sticks up through the soil and appears on the surface is known as an outcrop. Since the rocks may, if the soil is thin enough, appear at the surface in the areas where the geologic map shows the formation to be, patterns like the dendritic pattern are also called *outcrop patterns*.

Inclined Strata

A second possibility for rock structures is that the strata may remain parallel to each other, but be tilted, as in the bottom of Figure 2-3. Geologists call this situation inclined or *dipping beds;* looking at Figure 2-3, we would say the beds dip to the left (or away from us) since the beds go down into the ground as we go that way. Looking at the top drawing of Figure 2-3, we can see that the geologic map for this case shows parallel bands of color; that is the normal outcrop pattern or geologic map pattern for inclined strata. On a map of a real location, the bands may not be as evenly parallel, but the pattern is definitely different from the dendritic one.

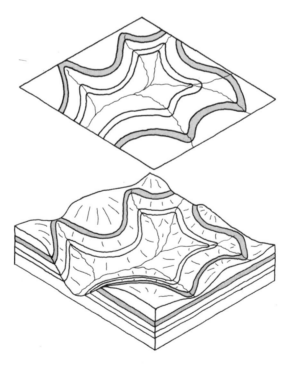

Fig. 2-2. A block diagram of horizontal strata, and a geologic map of the same area above it.

We can also gain another piece of information from the block and its map in Figure 2-3. Looking at the block, we can see that the bed which was at the bottom of this stack of strata before tilting occurred is the one labelled A, so it is the oldest bed. Then we go up and younger to B and C (the oldest one to show on the top of the block), and continue in sequence to layer G, which is the youngest of the beds. If we walked from C to G on the top of the block, we would be going "down-dip" because the beds dip left. We would be walking on younger and younger rocks as we moved from C to G. So, the general rule is "younger beds lie in the down-dip direction," and this applies correctly to our C-to-G case. The advantage to understanding the rule comes in looking at the map in Figure 2-3. First, the parallel bands on the map tell us the beds are inclined. If we then read the legend of our geologic map and find that C is older than G, we'd also know the beds dip toward G. And we can know that from looking only at the map—we don't need the block

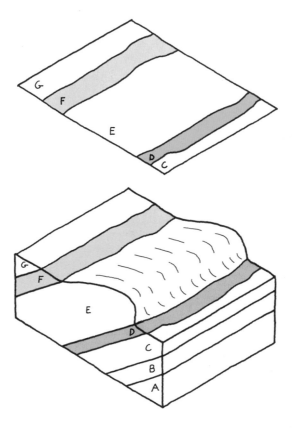

Fig. 2-3. Inclined strata and a geologic map of the same area.

diagram to go with it. We'll use this type of analysis later on to draw conclusions about Maryland rocks.

The block in Figure 2-3 also begins to show the relation between structure and topography. Though all the rocks were tilted up on one end, there is only one ridge across the top of the block. Presumably, the rock layer labelled E is more resistant than the ones around it, so it is left as the others wear away. The tilting causes all the layers to display the parallel band pattern, but their relative resistance finally determines which will be high and which will be low.

While we're looking at dipping rocks, we should cover some terminology that relates to them. The angle that the layers of rock make with a horizontal surface (and that surface may be the same as the ground) is simply called the *dip*. So, if a layer slants down into the ground at 20° below horizontal, then we'd say the dip is 20°. As the rock layer intersects the horizontal, it makes a line, just like the line where the floor and the wall of a room meet. For rocks, this line and its direction are called the *strike* of the rock layer. For the dipping rocks in Figure 2-3, the strike is parallel to the boundaries between adjacent colors on the geologic map. These terms are a little technical, but will come up again when we look at faults later in this chapter.

Folded Rocks

The next rock structures we'll look at are folds. The two basic types of folds are shown in Figure 2-4. The left side (2-4A) shows an up-fold called an anticline, while the right side (2-4B) shows a down-fold called a syncline. In the anticline, it is easy to see how the bed labelled C once connected across the top, and the block shows how it has eroded away after a time. Similarly, the bed labelled D on the syncline once extended farther up and out, but these parts of the D bed have eroded away.

The maps at the top of Figure 2-4 show parallel lines somewhat like the map for dipping beds we saw before. But notice how the maps for folds show repeating colors such that the same ones are on both sides of a single central color—the colors form *mirror images* on both sides of the central one. A glance back at Figure 2-3 shows no such repeating colors. This is how we can tell that a map shows folds—there are parallel bands of repeating formations in mirror images.

As with inclined beds, patterns of folds combined with knowledge of the age of the rock can be used to get more information from the map. For example, looking at the block diagram of the anticline, we can see that bed D is the youngest of the surface beds, and B is the oldest. Looking at the map, we see that B is the bed in the middle of the mirror-image pattern,

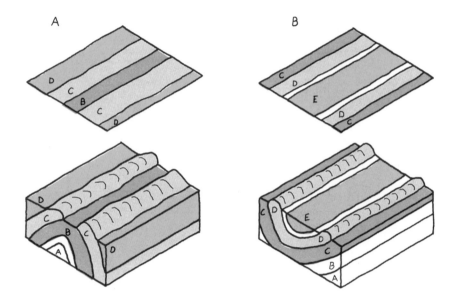

Fig. 2-4. Folds and their geologic maps: A. Anticline, B. Syncline.

which gives us a general principle: If the oldest band is in the middle of parallel bands of rock, the structure is an anticline. Knowing the principle, we can now recognize another anticline on a geologic map by noting the age patterns, without needing to see the view from the side. The opposite occurs with a syncline. In Figure 2-4B, E is the youngest bed and C is the oldest to appear on the surface. Therefore our pattern for synclines is that the youngest is in the middle, the reverse of an anticline. There are many folds like these in Maryland; the ridges on the block diagrams can give you a clue as to what province they are in.

Note that though the rocks in Figure 2-4A were bent up, and those in 2-4B were bent down, there are ridges in both cases. This is because both types of folding result in long narrow bands, or outcrops, on the surface. Then, regardless of which way the rock is inclined at the surface, resistant layers make ridges, and nonresistant ones make valleys. We'll look at this in more detail for Maryland in the next chapter.

Folds, however, can be more complicated than the ones in Figure 2-4. It is also possible that the fold itself can be tilted into the ground along its long direction; it is then said to be *plunging*. Folds of this type are shown in Figure 2-5, with 2-5A being a plunging anticline, while 2-5B is a plunging syncline. In both, the arrow labelled D shows the direction of plunge in three-dimensional space. Meanwhile, the arrow labelled P shows the direction of plunge on a horizontal plane, which is similar to the surface of the earth, and therefore like the geologic map. Notice the darkest color region in each diagram, showing the area of a particular rock unit that would be on the surface. That region makes a rounded V, but not like the one in horizontal rocks because there's no stream crossing the V. You can see that the V points in the direction of plunge (P) of the anticline (that is, the direction which is down for the overall fold), but the V points away from the plunge direction of the syncline.

A single fold seldom exists alone, since the side of an anticline where the rock layer is going down into the ground usually turns and comes back out again. In making the turn upward, a syncline is formed. It's easy to push the sides of the two blocks of Figure 2-4 together in your mind, so layers D of the anticline and C on the syncline become one continuous layer, and the two folds are side by side. If folds beside each other are also plunging, you get a zig-zag or S-shaped pattern, as shown in Figure 2-6. Again, this pattern becomes more obvious as resistant layers make ridges and nonresistant ones in between wear away. These folds are found in Maryland, too.

Domes and Basins—A final type of fold structure that we need to cover is the kind caused by pressure upward from a single point (or area), or sinking at a single point. Pressure upward creates a dome, in which the rock layers dip outward in all directions from the center, like a cloth draped over a raised finger. Pressure from below to create this could be caused, for example, by an injection of melted rock in existing rock layers. This is called an intrusion, and in some locations it domes up the layers above it. The opposite occurs

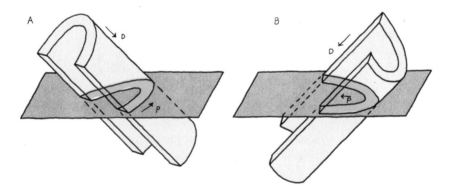

Fig. 2-5. Plunging folds: A. *Anticline,* B. *Syncline. In both,* D *is the direction of the plunge of the fold in space, while* P *is the direction of plunge on the horizontal surface.*

when a sinking or subsidence in one place creates a bowl-like structure called a basin, in which the layers all dip toward the center.

A newly made dome would probably be a hill or a mountain, and a recently created basin would be a valley. But erosion would cut off the top

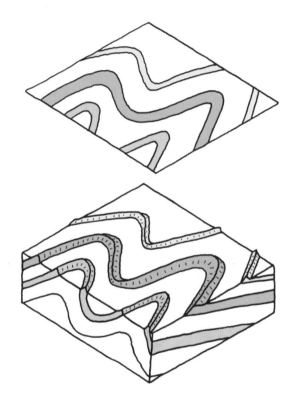

Fig. 2-6. Plunging folds and a geologic map of the same area.

of the dome, and wear down the edges of a basin. The result is that a geologic map shows the inside layers of the dome or basin, which make a roughly circular pattern like those seen in Figure 2-7. The basin is like the bull's-eye pattern made by the top edges of a set of mixing bowls nested inside each other; the dome is the same set turned upside down. Another analogy would be the circles made by layers of an onion which is cut into horizontal slices; whether it's a dome or a basin depends on whether the top or bottom half of the onion still remains. Domes and basins in the earth are recognizable circles in some locations, but also get stretched into elongated or other shapes. In that case, a dome might look like a loaf of Italian bread, and so have two long sides, while a basin might be canoe-shaped. However, on a geologic map of either of these, you can follow the color for a particular formation all the way around the structure; that would tell you it was a dome or a basin.

Given a circular pattern on the map, how do we tell if we're looking at a dome or a basin? Ages of the rocks again help answer our question here. A dome, like an anticline and for the same reasons, has the oldest rocks exposed in the center of the pattern. A basin is the opposite, with the youngest rocks in the center. Once you've determined which it is, you can mentally pull up on the center of a dome on a geologic map to "see" the underground structure. Similarly, you can mentally push down on the center of a basin as a reminder that the layers slope in toward the middle.

Remember that the shape of the rock structure does not by itself control the landforms. Figure 2-7A shows a dome that really is high in the middle, because the rock in the center must be resistant. But Figure 2-7B is also fairly high in the center, even though it's structurally a basin. This can happen if the rock in the center is more resistant than those around it.

By the way, many geologic maps also have at least one *cross section*. A cross section is the view you would get if you could cut a deep slice into the earth and then look at the rock layers from the side, instead of only from

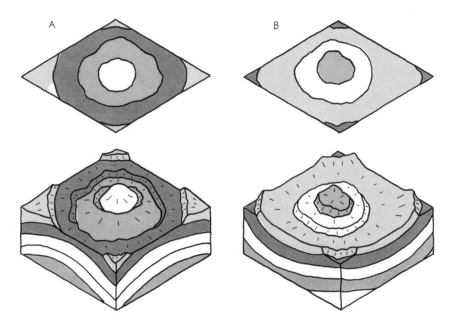

Fig. 2-7. Block diagrams and geologic maps: A. *Dome,* B. *Basin.*

the top. Thus, the cross section is like the *sides* of the blocks that we've been using to illustrate geologic structures—they make it easy to see folds because we can see the actual bends in the layers. There is also always a labelled line on the map which shows where the cut for the cross section was made; the view of the cross section is then always at right angles to that line. Often cross sections are vertically exaggerated; that is, they have been stretched up so they are taller than they would be if their vertical scale were the same as the map's scale. This is done so more details can be seen. The biggest effect of this enlargement is that it makes rock layers in the cross section look more tilted than they really are.

Faults

Faults are another general type of structure that occurs throughout Maryland. A fault is a break in the rock, like a crack in a block or brick wall, along which some motion has occurred. Both folds and faults are formed due to pressure in the earth. Whether a rock folds or breaks depends on factors like the rock type, the surrounding rocks, the amount of pressure, and how long the pressure is applied. When a rock does break, pressure on the fault causes motion along the fault surface, so rock layers can become offset from each other. The direction of motion and offset determines which of the two basic types of fault it is: strike-slip or dip-slip.

Strike-Slip Faults—In a strike-slip fault, the rocks move along the line, or strike, of the fault, as shown in Figure 2-8A. This means that the rocks slide past each other without moving up or down, like slipping two books by each other while they're lying on a table. The well-known San Andreas fault in California is of this type, but there are no major active strike-slip faults in Maryland. There are some faults that show evidence of strike-slip motion in the past, though the evidence is difficult to interpret.

Dip-Slip Faults—In a dip-slip fault, the rocks move up or down along the fault surface; after the motion, the rocks on one side are relatively higher than the same rocks on the other side of the fault, as in Figures 2-8B and 2-8C. Since dip-slip faults usually extend into the ground at an angle that isn't vertical (just like dipping rock strata), motion up or down also pushes the rocks either toward or away from each other. This is like pushing an object on a ramp: the object either goes up and toward the ramp, or down and away from the ramp. This type of motion separates the two kinds of dip-slip faults, as shown in Figure 2-9. Cross section A of that figure shows the situation before faulting, for comparison with what happens after the

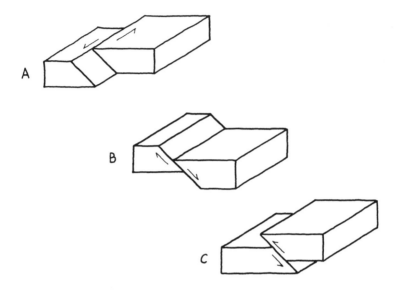

Fig. 2-8. Block diagrams of fault types: A. Strike-slip, B. Normal dip-slip, C. Reverse dip-slip.

fault motion. Cross section B shows a motion in which the rocks move apart, and so take up more length than they did before the movement. This results from a stretching, or tension, force in the earth and is called a *normal fault.* Cross section C shows a motion in which the rocks are pushed together, and so fit in less length than before. This is caused by a compression force, and is called a *reverse fault.* You can see that if we find a dip-slip fault in the earth and can determine what type it is, we will automatically know what type of forces existed when it formed: tension or compression. Though no dip-slip faults are currently known to be active in Maryland, there is ample evidence that they have occurred in the past.

One more subtype of fault is worth mentioning because it also appears in Maryland, and that is the *thrust fault.* A thrust fault is a reverse fault that has a low dip angle; that is, it's nearly horizontal. This means that compression (as we'd expect in a reverse fault) pushes one rock layer over the top of another, with the fault surface between them. Figure 2-10 is before-and-after drawings of a thrust fault; the layers pushed in this way are called thrust sheets. This can distort age relations: the top of Figure 2-10 would suggest A, B, and C are the same age, but the bottom drawing would make A seem older than C. As is obvious, thrust faulting considerably shortens the area covered by the rock units involved and moves some units a

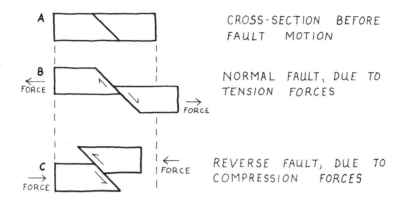

Fig. 2-9. Cross sections of dip-slip faults, showing lengthening and shortening effects.

substantial distance from where they started. Amazingly, there are examples in Maryland in which rocks have been pushed many miles along thrust faults, as we'll see in Chapter 5.

It is usual to have some folding occur near faults, as rocks frequently bend a bit before they break. But the extreme pressures that cause thrust faults may also cause extreme folds. Figure 2-11A shows such folds forming; by Figure 2-11B they have been pushed onto their sides, so they are lying down or *recumbent*. Continued pressure leads to the situation in Figure 2-11C, in which the recumbent fold has been broken off and pushed over other folds along one or more thrust faults. The resulting block of intensely

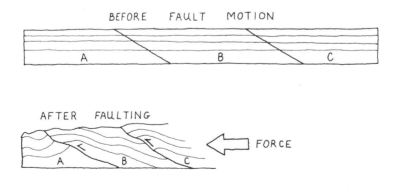

Fig. 2-10. Cross sections of thrust faulting, showing shortening and stacking of strata.

folded rock is called a *nappe*. We can see blocks like this on the sides of mountains in the European Alps today. We're also recognizing that the eroded remnants of them exist in Maryland, mostly recognizable by their thrust faults and strongly folded rocks.

Mapping Faults—Faults are usually shown on geologic maps by heavy black lines. At some faults on a map it's easy to see that the colors indicating the formations on each side of the fault are offset compared to each other; in other faults, especially thrust faults, this doesn't show very well. On some maps showing dip-slip faults like those we'll find in Maryland, beside the line of the fault you may find a *U* on one side and a *D* on the other. The U is on the side of the fault where the land moved up, and the D is on the down-thrown side. When you have a geologic map in front of you, the best way to decide if a fault is normal or reverse is to look at the arrows on the cross section of the map; you can then decide if the rocks have moved together or apart.

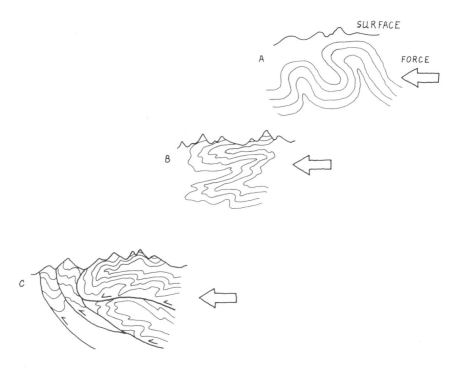

Fig. 2-11. Successive steps in the formation of a nappe.

There are, of course, many other geologic structures possible in addition to the ones mentioned above. However, this covers the major types we find in Maryland. With that introduction to structures and maps, we can now look at how weathering and the existing rock structures together produce the landforms of the state.

Chapter Three

How the Landscape of Maryland Was Made

We're now in a position to look at each of Maryland's physiographic provinces one at a time, to see how the rocks and structures of each province have influenced the landforms of the area. The coverage in Chapter 1 moved from east to west, but this time we'll start with the provinces that have the simplest structures, and then go on to more complex ones.

There are two ways we could examine the geologic structures of the provinces: One way is to describe the structures, then use the geologic maps to illustrate them. This is a bit like the procedure that geologists first investigating an area follow—they must examine and interpret the structures in the field, then represent them on a map. The second approach is to look at a geologic map and deduce from the map patterns what structures are shown. This is the method a geologist (or anyone else who can read a geologic map) would take if the area had already been mapped. Since that is the case in Maryland, we will use the second approach, which provides us the satisfaction of figuring out the geology based on the evidence in front of us.

A few words need to be said about the geologic maps of Maryland areas included in this book. First, there are many details that can't be included on maps this size. Relatively small, isolated areas of particular formations, intricate patterns of outcrops, or complex inter-fingerings of combinations of formations simply require larger maps to show up effectively. Also, two or more formations are often mapped as one unit to keep the patterns from being too complex to understand. If you want detailed map information for an area, the best sources are county and quadrangle geologic maps available from the Maryland or U.S. Geological Survey, whose addresses are listed in Appendix G. The maps in this book do, however, illustrate the general geology, plus some of the specifics, and will help you understand additional details to be found on large-scale maps.

The maps in this chapter were adapted from state, county, and quadrangle maps published by the Maryland and U.S. Geological Surveys. The

1968 state geologic map was the main source, with updates from newer maps and from information provided by geologists at the Maryland Geological Survey on areas currently under study. All the information on the maps was as accurate as possible at the time of publication, but may change as further study of the land gives us new facts that require different interpretations. The age of rock units can be found in Appendix E, as part of the geologic columns. Ages and descriptions of the rock types in each formation are in Appendix F.

Coastal Plain

Figure 3-1 is a geologic map of the Coastal Plain. The outcrop pattern on this map is recognizable as dendritic, especially in the lower western shore area. This indicates that the rocks are horizontal, or very nearly so. Reading the rock descriptions of the formations (Appendix F) reveals another fact: material below the soil of the Coastal Plain isn't solid rock at all, but mostly gravel, sand, and clay deposits. Geologists refer to these deposits as *unconsolidated sediments* because they have not been packed and cemented into hard sedimentary rocks.

It is not surprising, therefore, that this area has little relief, as erosion can easily remove any high spots that develop in these soft, uncemented materials. Also, as will be described in Chapter 5, these unconsolidated sediments have never been mountains, nor have they been subject to long erosion like the rest of the state. Instead, the Coastal Plain is the low area near the sea where eroded material has been *deposited,* catching everything that has washed down from the hills. These sediments accumulated in the nearly horizontal layers we find them in now. Parts of the Coastal Plain west of the Chesapeake Bay are higher and thus have more rolling hills than we find east of the Bay. But not much total relief has developed overall in the Coastal Plain because of its soft sediments and low elevation.

A few additional words may be useful in interpreting the geologic map. If we go down a stream in an area of horizontal strata, we should encounter older layers because we are getting lower in the rock column (Figure 2-2 may help make this clear). This shows up best when the streams cut deep canyons in the rock layers, which is clearly not the case in Maryland's Coastal Plain. However, we can still see examples of it in Figure 3-1, in the streams southeast of Washington, DC. If we start at the head of a small stream in an area of sediments of late Tertiary time (Pliocene) and go downstream toward the Potomac River (the streams are indicated by the dendritic pattern they create), we encounter older sediments, from early

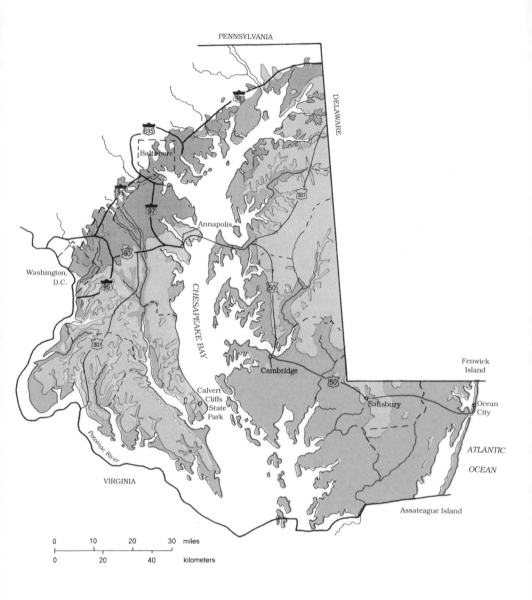

PENNSYLVANIA

DELAWARE

Baltimore

Washington,
D.C.

Annapolis

CHESAPEAKE BAY

Cambridge

Calvert
Cliffs
State
Park

Salisbury

Fenwick
Island

Ocean
City

ATLANTIC

OCEAN

Potomac River

VIRGINIA

Assateague Island

```
0        10       20        30   miles
├────────┼────────┼─────────┤
0            20           40      kilometers
```

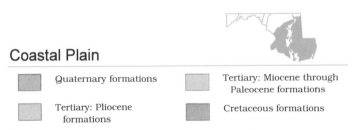

Coastal Plain

Quaternary formations	Tertiary: Miocene through Paleocene formations
Tertiary: Pliocene formations	Cretaceous formations

All white areas near the northwest edge of this map are metamorphic rocks of the Piedmont.

Tertiary time. Similarly, in the northeast corner of the Coastal Plain (Kent and southern Cecil counties), we go from sediments of late Tertiary time to ones of Cretaceous or early Tertiary time as we move downstream toward the Bay. This is what we would expect for horizontal strata.

However, if we continue to follow streams downhill in the Coastal Plain in the areas mentioned above, we often get to Quaternary, that is, *younger* sediments, near the Bay or river, which may seem incorrect. Actually, there's no problem: these Quaternary sediments are relatively recent deposits which simply cover over the older sediments. If we could dig down through the Quaternary deposits, we would find more of the older units we just found up the stream valleys. The Quaternary cover materials make up much of the immediate shoreline of the river estuaries and the Bay.

Just as the low relief of the land continues out into the Atlantic, so do the Coastal Plain sediment layers. The same nearly horizontal formations make up the Continental Shelf, and so it also is mostly sand, gravel, and clays.

The rocks of the Coastal Plain aren't precisely horizontal, however. This shows on the geologic map, though it takes a careful look to see it. Notice that the oldest Coastal Plain formations (Cretaceous age) occur in a band that runs southwest to northeast, between Washington, DC, and the northeast corner of Maryland. This band roughly parallels the Piedmont-Coastal Plain border. Looking for younger sediments, we can see that the major area of Tertiary sediments is in a broad band, again running southwest to northeast, across southern Maryland west of the Bay, and into the northern part of eastern Maryland. Still younger, the largest continuous area of Quaternary sediments of the Coastal Plain is the area southeast of Cambridge. The point of this is that sediments of the same age occur in parallel bands running southwest to northeast, and the bands are of younger sediments as we move southeast across the land. This is also shown in Appendix E-3, which shows that the oldest sediments of the Coastal Plain occur only on the western shore, while younger ones are only on the Eastern Shore. This is the pattern we saw for inclined beds, such as Figure 2-3. In the case of the Coastal Plain, the parallel bands we would expect for inclined strata are of varying widths, with indistinct and overlapping edges, but they are still apparent when we take a general view.

These bands mean the Coastal Plain sediment layers actually dip. The dip is at a very low angle, like a gentle slope, which is why we still get a dendritic pattern, and the parallel bands are hard to see. Since the beds get

Fig. 3-1. Geologic map of the Coastal Plain province.

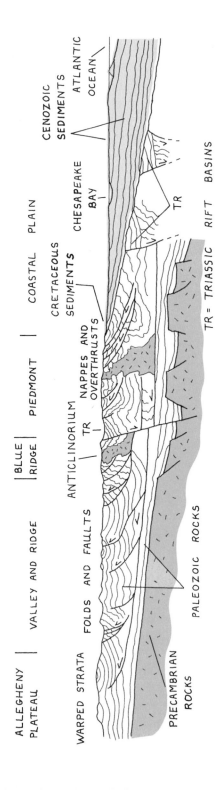

Fig. 3-2. Generalized geologic cross section of Maryland. This section conceptually follows the lines of Interstates 68 and 70, from the western end of the state to the Fall Zone, then goes southeast across the western and Eastern shores to the Atlantic. Since it is vertically exaggerated in order to show detail, the way some things look is misleading; for example, the dip on the Coastal Plain strata is really less steep than it looks. The same is true for the dip of all strata in the cross section. In addition, this is a generalized cross section, which means it shows the correct types of structures for each province, but does not attempt to show actual structures or their precise location because they would be out of proportion at this scale. To see cross sections of actual structures, look on a geologic map of the area of interest.

younger as we go south or southeast, towards the coast, the beds must be dipping in that direction since the rule we formed earlier was "younger beds lie in the down-dip direction." The proof for this comes in using the view that shows dip plainly: the cross section of the state, Figure 3-2. Looking at its eastern end, it is clear that the Coastal Plain rocks dip gently seaward—not much off horizontal, but a little. Notice that, as we would expect, the oldest strata are at the bottom of the stack of layers, and the youngest at the top. Thus, the oldest rocks are at the surface the farthest inland, and you cross the exposed edges of younger and younger layers as you go seaward.

Because all of these layers are composed of similar sediments, and no bed is particularly resistant, we don't notice many changes in landforms as we go across the region. There is often more relief on the (older) Tertiary and Cretaceous sediments, while the (younger) Quaternary ones are flatter, particularly because the latter sediments are so close to sea level anyway that not much relief can develop on them. In general, however, variations in resistance result in only minor variations in the landforms.

We should also look briefly at the barrier islands along the seacoast of the Coastal Plain, and how they form. They are made of sand, part of which is derived from the sediments that make up the Coastal Plain and part from sediments that are constantly being washed into the ocean by rivers that drain the Piedmont and the Appalachian areas. This river sediment is carried by ocean currents along the shore, mainly towards the south in this part of North America, and so the sediments get spread out to make long islands. Notice that this means these islands aren't static—material is always being washed away and new material added. If input equals output, the islands remain nearly the same, but this need not always be true. We'll come back to this situation in Chapter 6, but for now it is important to realize that the barrier islands are one of the most dynamic landforms in Maryland. Their creation and maintenance by sediments and currents is very complex, with many factors entering into the equation. They are constantly changing in small ways, and can occasionally change in big ways: barrier islands may even disappear entirely due to waves and currents from a large storm.

Allegheny Plateau

Moving on to another physiographic province, the next easiest to under-stand is the Allegheny Plateau. Figure 3-3 is the geologic map for this area. Using the formation names from the map key and the rock descriptions in

Appendix F, we can see, first, that all of the rocks of that area are sedimentary. Second, on the geologic map it's easy to see parallel bands in a mirror-image pattern, especially in the area between Oakland and Frostburg. This tells us the rocks are folded, and a look at the western end of the state cross section (Figure 3-2) shows this is correct. However, the cross section also shows the folds are gentle and broad, not tight; this is why the mirror-image pattern is easy to see on the map, because it results from wide, open folds. Geologists sometimes describe this kind of structure as *warped* layers, like a wooden board that has gentle bends, but is still basically flat. Since this is an area at a fairly high elevation, as a plateau should be, these rocks must be uplifted as well as warped. Thus, the overall view of the Plateau is layers of sedimentary rocks that have been folded slightly and uplifted.

To get a more detailed idea of the Plateau structure, it's helpful to follow a particular formation on the map and understand its three-dimensional shape. A good example is the Pottsville Group (mapped in Figure 3-3 with the Allegheny Group), since it appears all across the Plateau. We'll see what it does approximately along the line of Interstate 68. Starting in the east of the province, the Pottsville first appears several miles east of Frostburg. The structure around Frostburg is a syncline (youngest rock in the center), so the Pottsville must go under the town; it then comes up again a few miles west. The next structure to the west is an anticline (oldest rock in the center), which also agrees with our finding that the Pottsville is coming up out of the ground here, heading into the up-fold. The formation, now only in our imagination, goes up into the air and comes down again east of Grantsville. Of course, the rock layer actually did arch up like this once, but now it's eroded away. As we picture our layer on the way back down, it next passes under the Casselman River valley and comes up on the other side of this syncline, now west of Grantsville. Again heading up, it used to arch (though now it's eroded away) over the Keysers Ridge area but still comes down east of Friendsville. From there to the west the Pottsville is in a gentle syncline shape, close enough to the surface to have a few areas appear again in the northwest corner of the state. Overall, we can see that this formation and others above and below it make gentle wave structures across the Plateau region.

We can also see that the structure changes slightly in the northwestern half of Garrett County. Again using the geologic map, find the Pottsville Group where it occurs in two bands, one east and one west of Grantsville. If we trace these two bands (which are the sides of a syncline) south toward

Fig. 3-3. Geologic map of the Allegheny Plateaus province.

Allegheny Plateaus

Dunkard Group
Monongahela Group

Conemaugh Group

Allegheny Group
Pottsville Group

Mauch Chunk Formation
Greenbrier Formation
Purslane Formation
Rockwell Formation

Hampshire Formation

Foreknobs Formation
Scherr Formation
Braller Formation
Harrell Shale
Mahantango Formation
Marcellus Shale
Needmore Shale

Oriskany Sandstone
Shriver Chert

Helderberg Group
Tonoloway Limestone
Wills Creek Formation
Bloomsburg Formation
McKenzie Formation
Rochester Shale
Keefer Sandstone
Rose Hill Formation

Tuscarora Sandstone
Juniata Formation

PENNSYLVANIA

WEST VIRGINIA

WEST VIRGINIA

Cumberland

Frostburg

Lonaconing

Westernport

Savage River

Casselman River

Grantsville

Keysers Ridge

Deep Creek Lake

Accident

Oakland

Hutton

Friendsville

Youghiogheny River

Potomac River

40

68

Alt. 40

219

40

495

135

219

39

219

50

220

68

0 5 10 miles
0 5 10 15 kilometers

Deep Creek Lake, we find they meet, and so form a V. This means that the structure is a plunging syncline, and it plunges gently to the northeast. Looking farther southwest, we can see the Pottsville makes the same V pattern again (younger rocks in the center), but this time the V points the other way. That means the area north and west of Oakland is also a plunging syncline, but plunging to the southwest. So, structurally, the area from Grantsville to Hutton is like a roof gutter bent and propped up in the middle so each side slants down and away from the middle, which is located about at Deep Creek Lake. Note, however, that the land surface doesn't necessarily slope this way, just the rock layers.

A third structure found in the Allegheny Plateau province occurs in the area around Accident. On the geologic map, Accident lies within an island-like area of the Hampshire Formation, surrounded by the younger strata of the Mauch Chunk, Greenbrier, Purslane, and Rockwell formations. The boundary between these two map units is a closed loop that forms an elongated, indented circle. A roughly circular pattern like this, with the oldest rock in the center, indicates that this is nearly a dome, or we can also think of it as an anticline which plunges at both ends. So, the structure is like a long, narrow building with an arched roof, but the roof has been cut off at the top so we can look "inside" at the Hampshire formation.

You can find a similar elongated circle shape if you trace the Conemaugh Group around the Frostburg-to-Lonaconing area. This time, however, the rock in the middle, the Dunkard and Monongahela groups, is younger than the surrounding rock, and so this is a syncline. That makes the Frostburg-Lonaconing area a broad basin, or a syncline with turned-up ends (both ends plunge toward the center). The structure is like a broad, almost flat-bottomed boat shape. We can see from this and the previous example that the warped nature of the rock layers of the Plateau show bending in several directions at the same time.

Given these folds in the Allegheny Plateau, it is important to note that they are not the only causes for the high and low places of the area. As mentioned at the beginning of Chapter 2, the other factor that affects landforms (besides the structures we've just talked about) is weathering of the rocks. The sedimentary rocks of the Plateau differ significantly from each other in their resistance to weathering, with shale and siltstone being nonresistant, and sandstone usually resistant. This means that everywhere a resistant sandstone occurs at the surface, there is likely to be a high spot in the shape of that rock outcrop. So, we'll find a ridge if a long and narrow band of sandstone exists, or a generally hilly area if it's a broader shape. Similarly, we'll find a valley following the outcrop pattern of nonresistant rocks. Thus, though the structure may control the outcrop pattern of the rocks, their comparative resistances will control whether a high or low place develops there.

The formation that shows this best in the Plateau is, again, the Pottsville Group, since it contains several resistant sandstone members. This means that wherever we found the Pottsville when we looked before, we should also find a hill or mountain. And it doesn't matter how the rock layer is inclined as it reaches the surface, it should be high if sandstone is there. This idea works perfectly: a look at the topographic map of the Plateau (Figure 1-8) shows ridges in exactly the same positions as the Pottsville Formation on the geologic map. This is also shown in cross section in Figure 3-4. In the area west and south of Deep Creek Lake, streams have cut the ridges into separate hills, but they are still all the result of the Pottsville. There are some other resistant rocks that make ridges in the Plateau, especially the Purslane Formation, but the Pottsville is the cause of the major landforms. Often, the Pottsville serves as a cap over less resistant rocks—the ridge or mountain is in part made of softer rocks, but they are protected by the harder ones above.

One might well ask why a resistant layer folded into an anticline would erode away at its top, like the one in Figure 3-4 between Winding Ridge and Negro Mountain. After all, it's resistant, so why should it disappear? The answer is still being discussed by geomorphologists, but at least several factors seem to be involved. First, the folding tends to fracture the rock at the top of the anticline, just like the crack that starts at the top of a thick candy bar when you bend the ends down to split it in half. In anticlines, these fractures allow rainwater to penetrate into the rock layers, so there is increased weathering at the top of the anticline. Secondly, when the anticline is pushed up, the high area is subject to more severe weather, thus

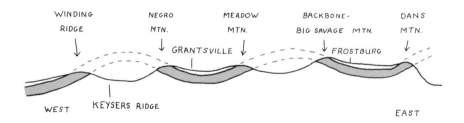

Fig. 3-4. Generalized geologic cross section of the Pottsville Group in the Allegheny Plateau in Maryland, roughly along the track of Interstate 68.

weathering is faster. Thirdly, there are slopes at the tops of the folds which contribute to landslides, and cause running water to erode more there than in the flatter valleys. All of these factors result in the top of the anticline being eroded away, exposing less resistant rocks within the fold. As these less resistant rocks erode, there is less support for the anticline around them. Thus, a *topographic inversion* develops, as shown in Figure 3-4, where the previously high anticlines have become valleys, while adjacent synclines are left as mountains.

Another landform that shows up occasionally is a *sinkhole*, which is a depression in the ground caused by the process of solution. Sinkholes form in limestone or marble, and are the result of these rocks dissolving away, especially around any cracks that have developed. Sinkholes that occur in Maryland are usually less than a few hundred feet across and a few tens of feet deep. Sometimes there is a hole in the bottom into which water flowing into the sinkhole disappears. Sinkholes can be found in the Plateau and in any of the Appalachian provinces, where the right rock type is at the surface. Though not large in a regional sense, sinkholes can dominate a piece of property, and are another landform directly caused by the underlying geology.

Valley and Ridge

Study of the Valley and Ridge is not too difficult once we've looked at the Allegheny Plateau, because we'll find a few of the same structures. However, some take on a new character, which is what makes the landforms different enough to classify them as a new province.

The geologic maps of the Valley and Ridge, Figures 3-5 and 3-6, show many parallel bands trending northeast-to-southwest across Maryland. The rock types are all sedimentary, and the mirror-image patterns tell us these are anticlines and synclines; all of these details are similar to what we found in the Plateau. But notice how the outcrop bands on the map are narrower in the Valley and Ridge. This thinness of the bands denotes tight folds, with layers sometimes so bent that we can only see their edges when we look straight down as the geologic map does. In fact, the distance between crests of folds in the Valley and Ridge is usually about half the distance between crests in the Plateau. In addition, the height of the folds in the Valley and Ridge tends to be roughly ten times the fold height in the Plateau. Thus, the Valley and Ridge rocks stand more on edge in contrast to the broad, nearly

Fig. 3-5. Geologic map of the western Valley and Ridge province.

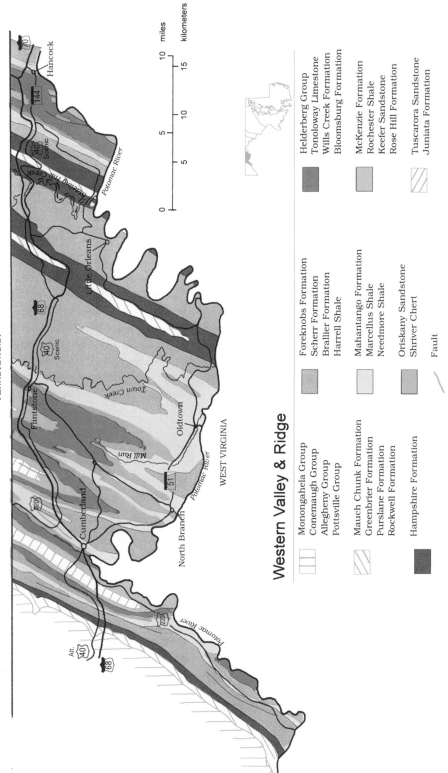

PENNSYLVANIA

Hancock

Scenic

Sideling Hill Creek

Potomac River

Little Orleans

Scenic

Flintstone

Town Creek

Oldtown

Mill Run

WEST VIRGINIA

Potomac River

Cumberland

North Branch

Potomac River

Alt.

0   5   10   miles
0   5   10   15   kilometers

Western Valley & Ridge

Monongahela Group
Conemaugh Group
Allegheny Group
Pottsville Group

Mauch Chunk Formation
Greenbrier Formation
Purslane Formation
Rockwell Formation

Hampshire Formation

Foreknobs Formation
Scherr Formation
Brallier Formation
Harrell Shale

Mahantango Formation
Marcellus Shale
Needmore Shale

Oriskany Sandstone
Shriver Chert

Fault

Helderberg Group
Tonoloway Limestone
Wills Creek Formation
Bloomsburg Formation

McKenzie Formation
Rochester Shale
Keefer Sandstone
Rose Hill Formation

Tuscarora Sandstone
Juniata Formation

horizontal folds of the Plateau. The tighter folds also show up on the cross section, Figure 3-2.

A close look at the folds of the Valley and Ridge shows they are often somewhat asymmetrical: anticlines tend to dip more gently on the southeast side, and more steeply on the northwest. This is shown on the geologic map by the bands of the same rock being wider to the southeast, and narrower to the northwest of an anticline. It's the opposite on synclines, of course, because the northwest limb of an anticline is the southeast limb of the adjacent syncline. In either case, this shape suggests rather intense pressure was put on the rocks to push the folds over this far.

Further evidence of strong folding is that some of the folds in the Valley and Ridge also are plunging. Several folds which show the characteristic V-shaped patterns appear in central Allegany County, and there are others in Washington County. This pattern occurs much more frequently in the Valley and Ridge than in the Plateau, and even becomes the zig-zag pattern (as in Figure 2-6) of adjacent plunging synclines and anticlines in a few places. Something has squeezed and bent these rock layers in several directions.

The forces that folded these rocks also produced faults, which can be seen on the geologic map throughout the Valley and Ridge province. Some faults result in offsets in the map patterns, while others do not; little visible offset occurs if the fault stays within a formation, or if it follows the boundary between two formations. Looking at the cross section, we can see that some of these faults have a low dip; that is, they're nearly horizontal at least over some part of their length. Therefore, these are thrust faults, and that tells us that the rock layers were squeezed together by horizontal forces sometime in the past.

The interaction between the structure and weathering in the Valley and Ridge is similar to what we saw in the Allegheny Plateau, but the folds express themselves more. Since the tight folds create outcrops in thin bands, the ridges and valleys are much narrower. So, the difference in weathering resistance between different formations is particularly emphasized due to the rapid changes in rock types as we move across the land. As a result, driving east to west across the state results in frequent climbs up mountains, and dips into valleys, as we go from one rock to the next, and then back to one we crossed before.

Fig. 3-6. Geologic map of the eastern Valley and Ridge Province.

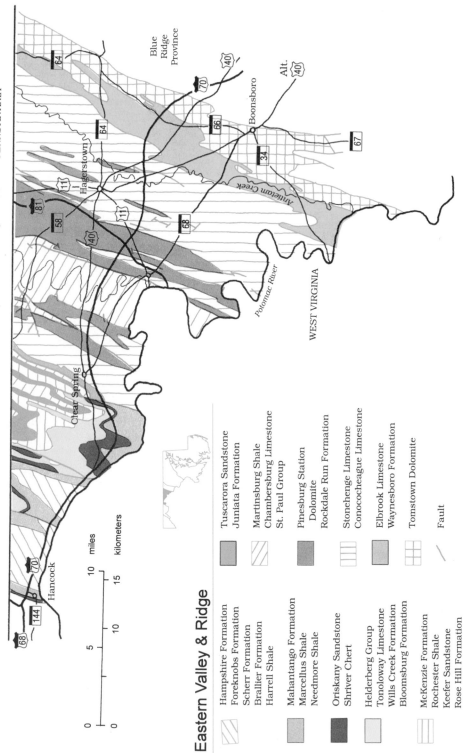

Eastern Valley & Ridge

PENNSYLVANIA

Conococheague Creek

Blue
Ridge
Province

Hagerstown

Boonsboro

Alt.

Clear Spring

Hancock

Antietam Creek

Potomac River

WEST VIRGINIA

miles

kilometers

	Hampshire Formation
	Foreknobs Formation
	Scherr Formation
	Brallier Formation
	Harrell Shale

	Mahantango Formation
	Marcellus Shale
	Needmore Shale

| | Oriskany Sandstone |
| | Shriver Chert |

	Helderberg Group
	Tonoloway Limestone
	Wills Creek Formation
	Bloomsburg Formation

	McKenzie Formation
	Rochester Shale
	Keefer Sandstone
	Rose Hill Formation

| | Tuscarora Sandstone |
| | Juniata Formation |

	Martinsburg Shale
	Chambersburg Limestone
	St. Paul Group

| | Pinesburg Station Dolomite |
| | Rockdale Run Formation |

| | Stonehenge Limestone |
| | Conococheague Limestone |

| | Elbrook Limestone |
| | Waynesboro Formation |

| | Tomstown Dolomite |

| | Fault |

How folded rocks develop into ridges on resistant rocks and valleys in nonresistant ones is shown in a general way in Figure 3-7. Now the pattern we began to see in the Plateau shows up even more clearly, and in a greater variety of situations. Notice in Figure 3-7 that, though an anticline is always an up-fold, the center of an eroded anticline could be either a ridge (A) or a valley (D) depending on the resistance of the rock exposed at the surface. Similarly, the down-fold of a syncline might be found either in a valley (B) or on top of a mountain (E), depending on the rocks involved. Other high spots can occur where the rocks are merely inclined (C). The interaction of structure and erosion is very apparent here.

Looking now at actual places in the Valley and Ridge, Figure 3-8 shows the major resistant rocks that create the mountains, and their structure in cross section. We can see from this that the recurrence of just three resistant sandstones is responsible for nearly all of the major ridges. The Tuscarora makes Haystack and Wills mountains, Evitts Mountain, and each of the Bear Pond mountains. The Purslane Formation makes Town Hill and Sideling Hill. The Oriskany Sandstone makes smaller ridges, but they are numerous: Fort Hill; Shriver Ridge; Irons, Collier, Martin, and Warrior

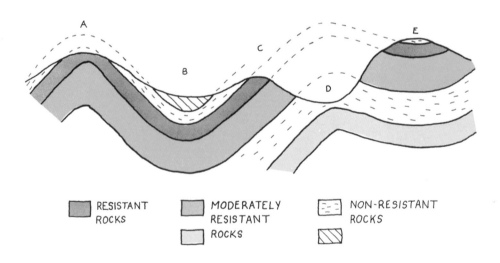

RESISTANT ROCKS   MODERATELY RESISTANT ROCKS   NON-RESISTANT ROCKS

Fig. 3-7. Geologic cross section showing how hills and valleys develop on folded rocks that have varying resistance to weathering.

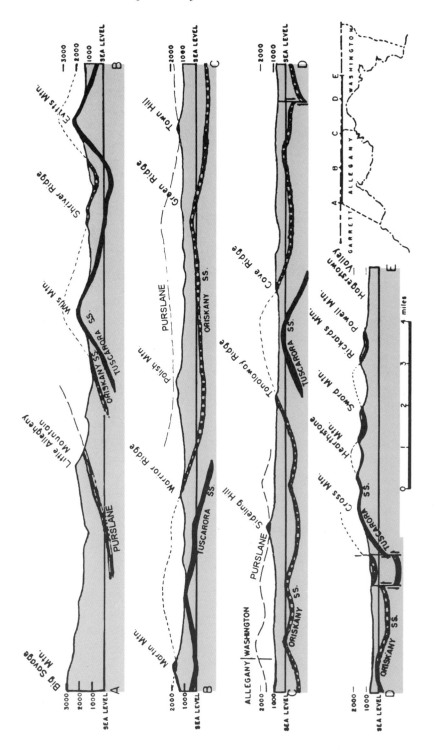

Fig. 3-8. Resistant rock layers of the Valley and Ridge province. Source: Maryland Geological Survey Bulletin 19, Geography and Geology of Maryland.

mountains; Stratford Ridge; Tonoloway Ridge; Roundtop Hill and Cove
Ridge; Elbow and Stone Quarry ridges; and Moore Knob. At each place on
the geologic map where one of these rocks is at the surface, the topographic
map shows a hill or mountain. A few other sandstone units create high spots
as well: The Foreknobs Formation supports Polish and Ragged mountains
and Green Ridge; and a sandstone plus conglomerate member in the
Mahantango Formation supports Coon and Orchard ridges.

Figure 3-8 shows some examples in which the land is like the underlying
structure, for instances, as the anticlines that make up Wills and Evitts
Mountains, or the synclines in the valleys between Hearthstone and Sword,
or Rickards and Powell mountains. We can also find cases where the
landforms are the opposite of the fold structure: the valley between Tonoloway and Cove ridges is the center of an anticline, and Martin Mountain,
Town Hill, and Sideling Hill are all synclines that are mountains. The last
example, Sideling Hill, is particularly notable because the large road cut on
Interstate 68 exposes the synclinal structure in a clear and spectacular
cross section. This is one of the best places in the state to see part of the
structure of the Appalachians, and is one of the best road cuts of geologic
interest in the entire eastern United States. The Exhibit Center there is
definitely worth a visit.

The structure of plunging folds also affects the topography of the Valley
and Ridge. For example, notice that there are several V-shaped patterns in
the geologic map just east of Cumberland. A check of their rock ages shows
there are two major anticlines here, with a syncline in the middle. The
Oriskany Sandstone is the major ridge-producer in this area; in the northern
part of the state, it makes Warrior Ridge and Martin Mountain, as shown in
Figure 3-8. To the south, the Oriskany creates mountains which also have
that pattern: Warrior Mountain on the east and Irons Mountain farther west
both are V-shaped at their southern ends. Between these mountains, the
two bands of Oriskany that occur (they're the center of small anticlines)
create Collier Mountain and Martin Mountain to the east. Though they're
small, all of these mountains are visible in the satellite photo, Figure 1-1.
An example in another area can be found in the plunging folds northwest
of Clear Spring. These folds make the steep, curving slopes of the Bear Pond
Mountains, visible in Figure 1-1 as the first set of ridges west of the
Hagerstown Valley, and including the area between Cross Mountain and
Powell Mountain in Figure 3-8. The point is that the mountains are faithful
reproductions of the bands of resistant rock regardless of their shape,
because that is why the mountains are there. The rocks are the cause, and
the mountains are the effect.

The Hagerstown Valley, which is Maryland's part of the Great Valley, has many examples of how the underlying rocks relate to the features we see. Most of the valley is underlain by limestones and shales, neither of which is very resistant to erosion—this is why the land has been eroded into a valley. The poorly resistant shales of the Martinsburg Formation have had a strong effect on the meandering path of Conococheague Creek, as the creek stays almost entirely in this formation. Also, the limestones of the valley dissolve underground to make caves in places, though only one, Crystal Grottoes south of Boonsboro, is open and easily accessible to the public. If you visit a cave such as this one, you will be able to see both the effects of the erosion that formed it, and the intricate and beautiful ways in which the calcite gets redeposited inside the cave—this is another place of geologic interest worth a visit. Finally, the muds that originally made the shales, plus clays mixed with the minerals of the limestones, contribute to the fertility of the farm soils in the Hagerstown Valley.

Because Maryland contains only a relatively narrow east-west band across the Valley and Ridge, this province seems confined to a small area. But it continues into the adjoining states (see Figure 1-2), and throughout its extent the province shows all of the features we've talked about for Maryland. The ridges are easily visible north and south of Maryland in the satellite photo, and some ridges show the V-shape of plunging folds. Also, if you travel to the Great Valley in Pennsylvania, Virginia, or West Virginia, you will find caves that have developed in the limestones there, just as they have in Maryland.

Blue Ridge

The Blue Ridge province in Maryland is dominated by a single large structure, though there are some complications that make it a bit difficult to see. If you look at the geologic map (Figure 3-9), you can determine the center of the Blue Ridge by finding the band of Loudoun/Catoctin/Swift Run formations that runs north-south between Boonsboro and Frederick. Looking east and west, you can see the Weverton Formation on both sides of the Catoctin Formation, and the Harpers Formation beyond that. This establishes our overall pattern: irregular parallel bands in mirror images. Since the oldest rock is in the center of the pattern, with the even-older Middletown Gneiss in the center of that to the south, the whole Blue Ridge must be an anticline.

Technically, the structure of the Blue Ridge is called an *anticlinorium,* which means it is an anticline overall, but with many smaller folds

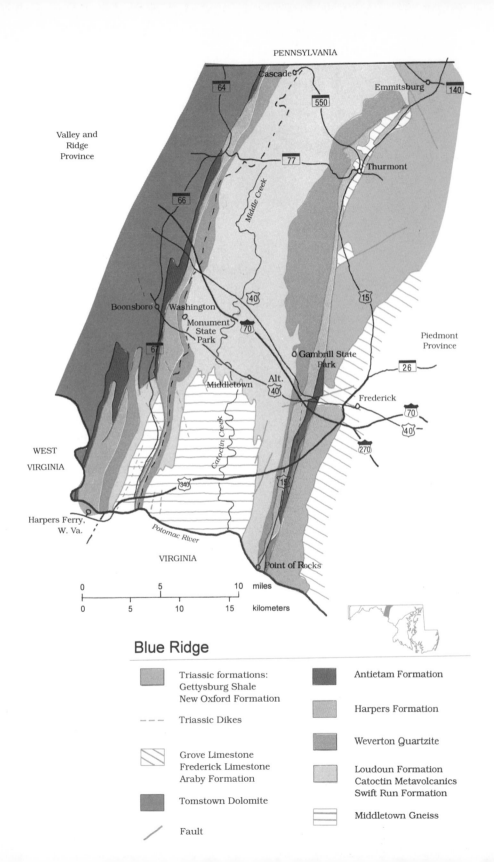

PENNSYLVANIA

Cascade

Emmitsburg

64

550

140

Valley and
Ridge
Province

77

Thurmont

Middle Creek

66

40

15

Boonsboro   Washington

Piedmont
Province

Monument
State
Park

70

67

Gambrill State
Park

26

Middletown

Alt.
40

Frederick

70

WEST

40

VIRGINIA

Catoctin Creek

270

340

15

Harpers Ferry,
W. Va.

Potomac River

VIRGINIA

Point of Rocks

0   5   10   miles

0   5   10   15   kilometers

Blue Ridge

Triassic formations:
Gettysburg Shale
New Oxford Formation

Antietam Formation

Triassic Dikes

Harpers Formation

Grove Limestone
Frederick Limestone
Araby Formation

Weverton Quartzite

Loudoun Formation
Catoctin Metavolcanics
Swift Run Formation

Tomstown Dolomite

Middletown Gneiss

Fault

throughout its structure as well. You could think of this like a sheet draped quickly over a bed—the sheet will go up one side and down the other overall, but will have other up-and-down wrinkles in it also. The west side of the Blue Ridge anticlinorium dips more steeply than the east side, and there are more faults and complex folds on the west side. Some of these folds and faults extend into the Tomstown Dolomite and other rock units in the Great Valley to the west of the Blue Ridge.

Most of the rock types in the Blue Ridge are noticeably different from what we've examined in the other provinces. With the exception of the Antietam Sandstone, the rocks in the Blue Ridge are metamorphic: phyllite, gneiss, metabasalt, and quartzite. The phyllite is not very widespread and so is not a major influence on the landforms. Each of the other rocks, however, does have an effect.

The gneiss is somewhat more resistant than the sedimentary rocks outside the Blue Ridge; therefore, much of the Middletown Valley is slightly higher than the Frederick and Hagerstown valleys on either side. The mountain-building events that have occurred in this area have fractured the rock, however, so streams do cut into it, creating the rolling topography of the valley. Though basalt might frequently be less resistant than gneiss, the metamorphism of the Catoctin Formation has made it into a hard, dense, sometimes slatelike resistant rock. Thus, the metabasalt forms the highlands of the northern Blue Ridge, and contributes to the ridges of South and Catoctin mountains.

Another rock which makes ridges is the Antietam Sandstone. Its slow weathering makes Red Hill, at the southwest edge of the province, and also makes the South Mountain ridge a bit wider northeast of Boonsboro. The hills in the Antietam are not as high as the main ridges, but are noticeable.

The rock that is the real standout, literally and figuratively, is the Weverton Formation. This quartzite is very hard and resistant, and is responsible for the highest places in the Blue Ridge. The locations of the mountains follow the locations of the Weverton Formation on the geologic map very well. Catoctin Mountain is the ridge resulting mainly from the quartzite band on the east side of the anticlinorium. This quartzite outcrop band and therefore Catoctin Mountain is narrow at the southern end at Point of Rocks, but broadens northwest of Frederick (see Figures 1-1 and 1-5 for the landforms). Meanwhile, the quartzite on the west side of the anticlinorium creates South Mountain, which remains narrow for its entire length in Maryland. Finally, the southwestern band of the Weverton makes

Fig. 3-9. Geologic map of the Blue Ridge province.

Elk Ridge. This ridge, where it meets the Potomac at Harpers Ferry, shows well the difficulty which weathering and erosion have in wearing down the Weverton: the cliffs above the narrowed river, and the rocks in the river itself, all indicate that the river cuts through these rocks more slowly than most other formations it crosses. People take advantage of the river's cutting action, and both roads and railroads cross South Mountain at this point. Farther north, both Interstate 70 and U.S. Route 40 cross South Mountain by taking advantage of a low spot caused by a gap in the band of the Weverton. Thus, the geology has a direct impact on people's actions in this area.

Piedmont

We've saved the Piedmont province for last because it is the most varied and complex. If we look at the Piedmont as a whole on the geologic map (Figure 3-10), its complexity is obvious, and no simple pattern emerges. The northeast-to-southwest trend of many outcrops continues to exist, but little else is immediately apparent overall. There are more faults here than in other provinces, but their patterns can be hard to interpret. We'll get more meaning out of looking at smaller areas, and we will cover each of these areas one at a time.

The Frederick Valley makes up the western edge of the Piedmont. The valley is located in the area underlain by limestones and dolomites, which are not particularly resistant. North of Frederick, we leave the Frederick Valley proper, but the area continues to be one of relatively low relief up to the Pennsylvania border. Though this area north of Frederick contains some sandstones and conglomerate, much of it is poorly resistant shale. So, the carbonates and the Triassic rocks of the western Piedmont make a more or less continuous valley from the Potomac to the Pennsylvania line. The more resistant rocks of the Blue Ridge to the west and the rest of the Piedmont to the east generally rise above this valley. Because of its higher ground, the Piedmont east of the Frederick Valley is often called the Piedmont Plateau.

Also in the Frederick Valley, a number of *dikes* are shown in Figure 3-10. A dike is a structure usually created by melted rock (magma) moving into cracks in existing rock, below the surface. In this case, the dikes intruded into the cracks caused by the formation of the rift valley (described in Chapter 5). The magma solidified into a rock called diabase, which is similar to basalt. These dikes do not have a major effect on the landforms of the province, though some of the thicker ones do produce low ridges. Also, farm machinery can get damaged by plowing over the resistant rocks, so the ground above the dikes may be left in a wooded condition.

A different part of the Piedmont lies in the area east of Frederick and west of Baltimore, and continues north and slightly east of Baltimore. As we can see from the geologic map and rock descriptions, most of the rocks in this area are metamorphic. That tells us they have been subjected to high heat and pressure deep within the earth's crust; that pressure has also folded and faulted these rocks rather intensely. The results of the folding and thrust faulting are so complicated and intricate that it is difficult to sort them out. An appreciation of the problem might be gained by an analogy: Suppose you were looking at a part of a bedsheet sticking out of the top of a pile of clothes in a clothes drier that had just stopped turning. It would be hard to determine exactly how the sheet was folded or twisted on the *inside* of the pile, just by looking at the top. The rocks of the Piedmont haven't been tumbled quite as much as clothes in a drier, but they have "been through the wringer" and are very twisted and pushed over each other.

However, geologists can figure out complex and intricate folds if enough information is available, which points to the second obstacle to under-standing the Piedmont: there are few good outcrops to study, due to the soil cover, vegetation, and people's changes to the land. As a result, it takes a lot of work to find out in detail what's there, and this work is still in progress. Because geologic maps in this area are currently being updated, the rocks of the west-central Piedmont are grouped and generalized in Figure 3-10. Within each mapped group, there are several separate areas of each formation, showing folds and faults. For now, we'll leave the structure description as "complex" for this area and see if rock types will help us understand the landforms.

Most of the metamorphic rocks in the Piedmont are similar in their resistance to erosion, thus no one rock unit stands out to the extent we've seen in other provinces. A few lines of hills are visible in the satellite picture (Figure 1-1), but the rolling hills are rather uniform over the area in general. Streams cut fairly steep valleys in a good many areas, and are quite apparent when driving around the Piedmont. But the relief is usually only around 200 feet at most, which is much less than that of provinces farther west.

The relief does increase in a few places in the Piedmont. One such place is Sugarloaf Mountain, and the reason it exists is that it's made of quartzite. Like the quartzite of the Blue Ridge, this rock is noticeably more resistant than the other Piedmont rocks, leaving Sugarloaf Mountain high, as other rocks around it erode away. Another high relief area is at Rocks State Park, created by a small band of metaconglomerate and quartzite. Similar rocks plus the dense Peach Bottom Slate hold up the hills in north-central Harford County.

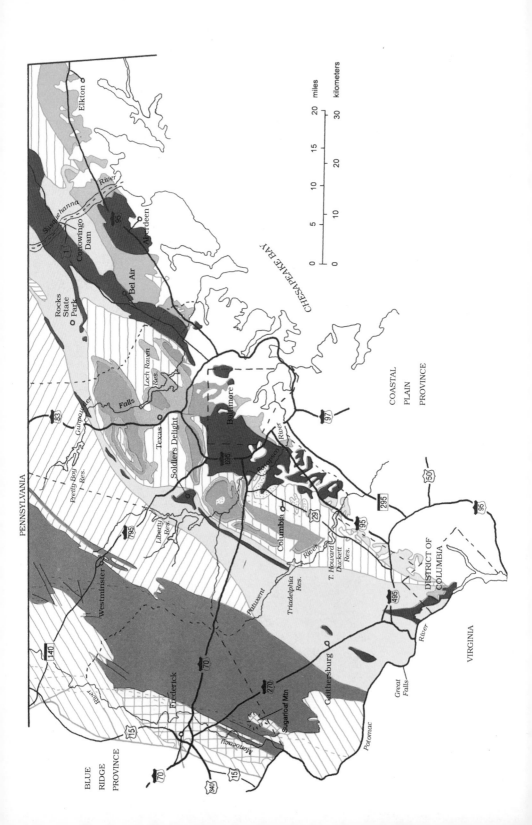

Piedmont

 Triassic formations:
Gettysburg Shale
New Oxford Formation

- - - Triassic Dikes

 Grove Limestone
Frederick Limestone

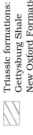 Marburg Formation
Cash Smith Formation
Araby Formation
Ijamsville Formation
Urbana Formation
Linganore Nappe:*
Gillis Formation
Sams Creek Formation

 Sugarloaf Mtn. Quartzite

 Peach Bottom Slate
Cardiff Metacon-
glomerate*

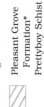 Pleasant Grove
Formation*
Prettyboy Schist

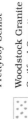 Woodstock Granite
Guilford Granite
Ellicott City Granite

 Oella*
Loch Raven (& similar schists)

 Cockeysville Marble
Setters Formation

 Baltimore Gneiss

 Sykesville Formation*
Morgan Run Formation*
Conowingo Diamictite*

 James Run–Port
Deposit Thrust Sheet

Baltimore Gabbro
Complex*
Aberdeen Metagabbro*
Other mafic & ultramafic
rocks*

 Fault

All white areas near the southeast edge of this map are sediments of the
Coastal Plain, Cretaceous or Cenozoic in age.
* Boundaries between these units and adjacent Piedmont units are usually
faults.

Fig. 3-10. Geologic map of the Piedmont province.

If we take a closer look at the Piedmont's metamorphic area, some other differences in rock weathering show up. For example, the Cockeysville Marble is noticeably less resistant than the other rocks in the area. Therefore, at most places where marble occurs, there is a valley. As shown on the geologic map, the largest continuous area of marble is north of Baltimore, where it underlies several interconnected valleys. These include the Greenspring and Worthington valleys, and the route of I-83 between them; they are given the name Timonium Valley overall.

Our closer look also reveals some structures that we can recognize in the rocks. North and west of Baltimore are rounded, often elongated occurrences of Baltimore Gneiss. Several of these are surrounded, or nearly so, by a ring, mapped in Figure 3-10 as the Cockeysville and Setters formations. More detailed maps would show that the Setters is usually in contact with the Baltimore Gneiss, and then the younger marble is beyond, farther from the gneiss. Continuing away from the gneiss centers, we find the marble is surrounded by the even younger Loch Raven and Oella rocks. These are rough circles, with the oldest rock in the center, and therefore are domes. But continuing analysis of these structures suggests that some of them actually aren't simple domes after all. Instead, they are probably nappes, which are anticlines lying on their side which have been further folded and faulted. Erosion of these distorted, squeezed layers has exposed their folds as domelike structures.

Like the folds we have seen in other provinces, these "domes" are not necessarily high in the middle just because they are up-warps in the rock. Many of these areas turn out to have higher elevations because the gneiss in the center of them is relatively resistant. The Setters Quartzite outside of the gneiss is very resistant, and usually makes a low steep ridge or hill which is particularly noticeable because it lies adjacent to the less resistant Cockeysville Marble. Thus, the marble valleys (the various areas of the Timonium Valley, mentioned above) may seem especially deep when you're travelling through them because of the large relief to the quartzite ridges. So, on the ground, the differences in rock resistance are quite apparent, and the landforms accurately reflect the outcrop pattern of the rocks and the geologic map. However, the dome structures can be found only with difficulty on the satellite photo because their landform effect is not as large as the effect of the folds in provinces farther west. Also, the dome-shaped folds rarely create more relief than is found in the other Piedmont Plateau areas, where the rock is often folded as well.

The last area of the Piedmont is the northeast corner, where the geologic map shows a variety of rock types: sedimentary, volcanic, and intrusive rocks, which have all been metamorphosed, and then folded and thrust faulted into outcrop bands that run northeast to southwest. There are rolling hills here but few prominent ridges, so the resistance of these rocks is fairly uniform. The elevation of the area, however, becomes especially noticeable at the Susquehanna River, and the hard rocks of the area are able to maintain steep bluffs above the river. The rocks and islands left in the river below Conowingo Dam show how the Susquehanna has to cut through rather resistant bedrock to form its channel.

There is another structure which probably runs from the Piedmont to the Allegheny Plateau, but is not visible at the surface of the earth. Geologists have recently (mostly in the 1980s) found evidence that perhaps all the rocks in these provinces weren't formed where they currently are, but rather formed much farther east and have been pushed west over other similar rocks along a deep thrust fault. This is visible near the bottom of the geologic cross section, Figure 3-2. Though this is like a usual thrust fault, it is on a very large scale: a piece of land that is 100 or more miles wide is being pushed over another piece of continent. In a case like this, the top layer especially gets folded in the process of the move, and some of the folds we see in Maryland rocks may have formed in this way. The deep fault has no apparent direct effect on surface landforms, but if it exists, it is an important structure for understanding how Maryland was made.

The studies suggesting the existence of this deep fault weren't done in Maryland, but in similar Appalachian areas to the north and south; many geologists feel that the fault therefore also exists in Maryland. This fault and how it came to be is an area of continuing research by geologists, and new discoveries and interpretations relating to it will probably be made in the future. Though we know some things, this is an example of an area where we still have much to learn about the earth.

Piedmont-Coastal Zone Boundary: The Fall Zone

Our coverage of the last few provinces has moved east, and so that gets us conceptually back to the boundary between the Piedmont and the Coastal Plain. Remember from Chapter 1 that this boundary is also called the Fall Zone, and now that we've looked at the rocks of both provinces, we can see why the Fall Zone exists. The Piedmont is made of rather hard and resistant metamorphic rocks near the boundary, and the Coastal Plain is unconsolidated sand and gravel. Rivers flowing east to the Atlantic Ocean across this boundary have the same effect as water running off a road or driveway onto mud—the mud is easily eroded away while the pavement remains.

Eventually, there's a significant drop from the pavement to the bottom of the gully that begins to form as the mud washes away. When this happens in a river, you have a waterfall, where the land underlying the river changes from hard to soft rocks. This occurs at each stream along the boundary that separates the Piedmont from the Coastal Plain, and thus the Fall Zone is created.

It should be noted that some falls of the Fall Zone, such as Great Falls near Washington, DC, are no longer actually at the province boundary because the river still does erode even the harder rock of the Piedmont. This moves the falls farther upriver than the real province boundary, so they are now surrounded by metamorphic rock alone. In the case of Great Falls, erosion was helped along by the river running along a fault, which weakened the metamorphic rock and made it easier to weather and be removed. Here again, the structure explains the landforms we see.

It's also clear from the geologic maps why the boundary between the Coastal Plain and Piedmont is not a simple line. Along the southeast edge of Figure 3-10, patches of Coastal Plain sediments appear in the metamorphic rock areas. On the northwest edge of Figure 3-1, areas of Piedmont rocks appear in places surrounded by Coastal Plain materials. The cause of this patchwork appearance is not hard to understand. A look at Figure 3-2 shows that the Coastal Plain sediments get thinner as we go toward the Piedmont. At the boundary, the Coastal Plain sediments will be down to only a thin coating over the top of the metamorphic rocks below. It will be easy for erosion to remove this coating in some areas of the Coastal Plain, and let the Piedmont rocks show at the surface. You get a picture of this if you spread the fingers of your right hand and lay them over your left hand: you can see your left hand through the gaps between the fingers of your right hand. Meanwhile, some places in the Piedmont where the overlying Coastal Plain sediments have nearly completely worn away may have small areas still covered, which will show up as isolated patches. So, in the vicinity of the boundary, we may find either type of rock as we travel to different locations, and we have to settle for an *area* which is the boundary, not a single line.

This also is an example of how one geologic province may merge or overlap into another. The metamorphic rocks of the Piedmont actually continue south and east in Maryland below the Coastal Plain sediments—we know this from drilling records and other data collection methods. The metamorphic rocks are a mile or more below the surface by the time we get as far east as the Atlantic coast, but they are still there. Like other deeply buried structures, they have no direct effect on landforms, but do help us understand the geologic history of the state, as examined in Chapter 5.

That completes our look at how the rocks and their weathering create the Maryland landscape. You can now appreciate the variety of structures shown in the geologic cross section, Figure 3-2. Reading from east to west, they are: the gently dipping sediments of the Coastal Plain; the complex folding and faulting of the Piedmont; the quartzite ridges of the Blue Ridge anticlinal structure; the tight folds with faults of the Valley and Ridge; and the gently folded high layers of the Plateau. The variety of rock types, and indications of tremendous forces that folded and altered them, suggests there have been lots of changes in Maryland over geologic time. In Chapters 4 and 5, we will examine how these rocks and their structures came to be, during the last billion years of the earth's history.

Chapter Four

Determining Maryland's Geologic History

When geologists work on determining the history of the earth, they are doing detective work that Sherlock Holmes would be proud of. There are no people to ask or books to read when geologists are investigating the earth's history—everything must be deduced from physical evidence. Each rock and each geologic structure has a story to tell, but learning to read the message in the rocks takes knowledge and practice. If we are to understand the geologic history of Maryland, it will help to see the connection between the rocks of Maryland and their meanings in the story. So, the first part of this chapter will introduce how we learn about an area's history, and the second part will cover how the earth changes over time through plate tectonics. Then we can see how Maryland came to be the way it is in Chapter 5. As before, if you don't need the general geology principles, skip to the next chapter. If the introduction will help you, then the next two sections will be a sort of detective school for historical geology.

Principles of Historical Geology

Our understanding of how the earth came to be the way it is has changed over time. For a long time, it was assumed that everything was created or formed all at once, or in a short time. Any changes since then had been caused by sudden events such as volcanoes or earthquakes; named after these events, this doctrine is called catastrophism. An alternative theory called uniformitarianism is closer to what is believed by science today. This principle was proposed in 1785 by a Scottish scientist named James Hutton, and widely supported when Charles Lyell drew together many supporting facts for it in 1830. Uniformitarianism states that the geological processes we see going on around us today are the same as the ones that went on in the past, which created the geologic structures we see. Though the processes may seem to act too slowly to have been responsible for all the varied geology that we see on the earth, given enough time they can do anything. For example, slow erosion that results in virtually no visible changes from day to day can still eventually wear away entire mountain ranges. Uniformitarians did see that sudden events also change the earth, but said these events and other processes are cyclical and occur at the same rate in the present as in the past.

Today, scientists agree with a principle of modified uniformitarianism called actualism. This doctrine says that the same physical and chemical laws exist today as throughout geologic time, though we now recognize that the processes do not work at constant rates nor are they truly cyclical. Instead, the processes vary some over time and distance, and sudden events do make their mark in the geologic record. Further, as the earth changes, the processes have new situations to work on, and so produce slightly different results instead of returning to previous ones in cycles. So, the earth evolves, in a one-way direction, through a continuation of the same mechanisms at work. Overall, this means we can look at the processes going on around us today, and draw conclusions about what must have formed the structures which were created in the past.

Reading History from Rocks
The first thing that rocks can tell us about their history is the events that occurred or the conditions that existed when they formed. This can be determined in large measure by looking at each individual rock type or structure, without worrying too much (yet) about what came before or after it. Of the three major rock types, sedimentary rocks often provide the most details of their history, so we'll look at them first.

Sedimentary Rocks—Many sedimentary rocks are formed from material washed down from a continent, especially from mountainous areas, which is then deposited along a river crossing the continent, or in the ocean where the river ends. River sediments are called alluvial deposits, or simply *alluvium*. As the sediments flow down from high areas, they are deposited wherever the river slows down. Sometimes this occurs as a fast-flowing stream enters a flatter valley, possibly creating a broad deposit shaped like part of a cone and called an *alluvial fan*. Along streams and rivers moving across fairly flat land, alluvial deposits are common, as channels move around and floods leave behind sediments. These deposits create what is called a *floodplain* around the river; as floodplains broaden, they become known as *alluvial plains*.

Many aspects of alluvial deposits tell us something of their history, but we'll concentrate on three characteristics: First, we note that the river can carry away the smaller particles of silt and mud, but leaves behind the larger pieces of rock. This means that the deposits with the largest boulders or pebbles—that is, the coarser sediments—occur nearest the mountains. Also contributing to the pattern is the way the particles grind on each other as they roll downstream, so all the particles get smaller as they go downriver. Second, because it is hard for the river to carry this coarse material, the sediments will be in thicker layers close to the mountains, and in thinner layers farther down. Third, the deposits containing the most chemically unstable minerals (especially the mafic ones) tend to be found close to the mountains. Sediments containing these unstable minerals are called *immature sediments* because their weathering time has been short. During the trip down the river, these unstable minerals will have broken down and disappeared, leaving only resistant quartz sand, mixed with products of weathering like clay; these are called *mature sediments*. Using rock examples, we would say conglomerate and graywacke are of immature sediments which were probably deposited near their source, while shale and pure quartz sandstone would be more mature, and probably traveled some distance from their source before being deposited.

These factors mean that if we can trace a group of alluvial deposits and determine where they are coarser, thicker, and more immature, then we'll know where there was high land in the past, even if that land is no longer high. Summarized in Figure 4-1, this area of deposits is known as a *clastic wedge*, because it is wedge-shaped and made up of eroded particles which geologists classify generally as *clasts*. The sediments in the clastic wedge that are carried by rivers and deposited on land often contain a lot of red oxidized iron, since they have been exposed to both air and water. The resulting layers of rock are called *red beds*. Though red beds form in a variety of environments and the exact mechanisms of their formation are debated by geologists, a red color along with other evidence can be a clue to an alluvial environment at the time those sediments were deposited. (By the way, Figure 4-1 is, once again, vertically exaggerated so its details can be seen ; the wedge of sediments is actually much longer and thinner than it seems here.)

Most of the sediments that rivers carry eventually reach the ocean and are deposited there when the river current ends. These sediments are usually in the form of mud and sand, along with some gravel, by the time they get that far downstream. At the mouth of the river, the sediments pile up to make a *delta*, which covers a large area underwater and may also appear as low land where the river meets the sea. Sediments that were deposited in deltas appear often in the rock record, indicating that river processes in the past were the same as those we see today.

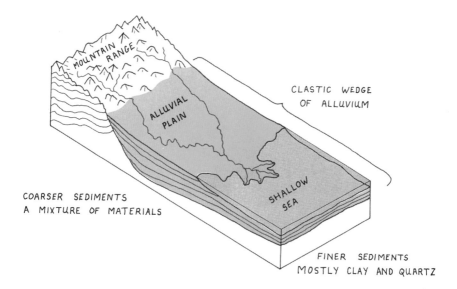

CLASTIC WEDGE
OF ALLUVIUM

MOUNTAIN RANGE

ALLUVIAL PLAIN

COARSER SEDIMENTS
A MIXTURE OF MATERIALS

SHALLOW SEA

FINER SEDIMENTS
MOSTLY CLAY AND QUARTZ

Fig. 4-1. Block diagram of a clastic wedge of sediments eroded from a mountain range.

The sediments at a delta are in part spread along the entire continental shore by ocean currents. A cross section of the types of sediments found as we go from the shore out to sea takes on certain typical distributions of sediment types, as shown in Figure 4-2. In 4-2A, pebbles are deposited near shore, where waves can carry them and move them around. Farther out, the water movement can still carry sand, but not pebbles. Still more distant from the shore, waves can carry only the finest grains of sediment, so the deposits are clay or mud. Notice that the change from one sediment type to another is gradual, with "inter-fingers," and covers a broad area.

Figure 4-2B shows another common situation. Here, sand is deposited by waves on a beach, and mud is carried farther out. Beyond, life in the sea may have an effect: organisms which concentrate calcite for shells and skeletons can create an accumulation of this material. Calcite usually forms in shallow, warm water where sunlight filters in to support plant life and a complex food chain of animals, such as happens on coral reefs. Beyond the reef there may be more clay deposits as the water deepens and wave action at depth is less. Of course, many other shoreline situations exist besides those shown in Figure 4-2A, and they are often more complex and changing as we travel along the shore. But this allows us to see how the material deposited depends on the environment in each location.

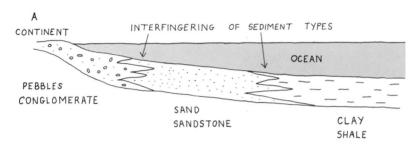

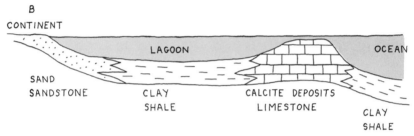

Fig. 4-2. Cross sections of two environments of deposition, showing the sediment types at different depths and the resulting rocks.

Given the sediments in Figure 4-2, we can see how the different rock types also shown in the drawing result. If we find these rocks beside each other in similar relationships, we have a good idea of where shorelines, reefs, and deeper water environments were, at the time when the rocks formed. Clearly, this is a powerful tool for determining the history of an area, because it helps us find the edges of continents in the past.

Another shoreline sedimentary deposit which tells us about the environment in which it formed is the *carbonate bank*. In the rock record, a carbonate bank consists of a thick accumulation of layers of limestone and dolomite, and covers a significant area. It forms where there is a relatively flat, broad shelf of land covered by shallow water along a shoreline. As mentioned above, those conditions often result in a good environment for organisms which contribute to limestone deposition. If the land subsides as the limestone accumulates, hundreds of feet of carbonate rock may be built up, even though the water stays about the same depth. This situation is occurring today around parts of Florida and in the Caribbean. When we find this in the rock record, we know an area had shallow warm water and rather stable conditions for a significant period of time.

Deposits also accumulate in deeper ocean water. Sometimes the material is mud which turns to shale. At other times and places, it is limestone created by the slow but steady rain of dead plant and animal life that falls to the bottom of the ocean. Either of these may be black or at least dark in color because of the low levels of oxygen found in deep water.

The flood plain and various watery environments mentioned above are only two of many types. Distinctive sediments, and therefore characteristic rocks, emerge from other environments, such as deserts or glaciers. But most of the sedimentary rocks in Maryland are from alluvial and fresh- or saltwater (marine) deposits.

Igneous Rocks—Geologists also use igneous rocks to determine past environments and events. They begin with simply observing the characteristics which allow them to name the rock, as summarized in Appendix B. A fine-grained igneous rock is one that cooled relatively quickly, and so probably formed from magma that flowed out of the ground as lava. Therefore, in the past, there was a volcano or other lava vent at the location of the rock. The rock type tells us what type of volcano it was, for example, basaltic or rhyolitic. Since the theory of plate tectonics (discussed later in this chapter) describes how different types of volcanoes form, geologists can conclude what was likely to be happening geologically to make this volcano in the past. Of course, they would go on to look for supporting evidence for that conclusion, but even the rock type alone is a good start in determining the history of a volcanic rock.

Alternatively, a coarse-grained igneous rock was probably magma squeezed into existing rock that cooled relatively slowly underground, resulting in an intrusive igneous rock. Intrusions occur below volcanoes, and so may be associated with the fine-grained rocks mentioned above. But intrusions, especially of granite, also can occur as part of an *orogeny*, which is the geologic name for a mountain-building event. The cause of orogenies is covered later in this chapter in the plate tectonics section, but for now we can note that finding granite is cause for geologists to look for further evidence of an orogeny. A *dike,* which is an intrusion of magma into a crack in the rock, suggests some forces existed to cause the crack; this might be below a volcano, or be further evidence of an orogeny. Thus, intrusions in various forms can also tell us about an area's history.

Beyond basically identifying characteristics of igneous rocks, geologists also examine them microscopically, and determine their chemical composition. This yields further information about the origin of the rocks, but a discussion of the factors involved in this type of study is beyond the scope of this book. However, facts that geologists have learned from such studies of Maryland's igneous rocks will help shape the history covered in Chapter 5.

Metamorphic Rocks—Metamorphic rocks tell us several things about their history. First, from the current rock type, we can generally conclude what the *protolith* was, that is, what type of rock it was before metamorphism. For example, slate was probably shale before metamorphism, quartzite was sandstone, and marble was limestone. Knowing the sedimentary protolith, geologists can use their knowledge of sedimentary rocks to determine the original environment in which the rock formed.

Second, we know that a metamorphic rock has had enough pressure on it to cause a change. That means it probably was buried deeply in the earth by other materials accumulating on top of it; finding a metamorphic rock means geologists must include a long period of time in their history of that area, to allow enough sediments to collect. The pressure to metamorphose rocks frequently also comes from a horizontal direction; such pressure is caused by the forces that cause orogenies. An orogeny will usually affect a wide area, so regions of metamorphic rock (such as the Piedmont) are excellent evidence that a mountain range formed. Thus, geologists have to include an orogeny in a history of the rock, and again, account for a long period of time—time for a mountain range to build and then wear away to expose the metamorphic rock at the surface.

Finally, because the rock type and its minerals change as more metamorphism occurs, scientists can get an idea of how much pressure, or how high a temperature, the rock has experienced. This indicates the intensity of the metamorphic event, suggesting, for example, how large an orogeny was. The change in the minerals and other aspects of the texture of the rock can even allow geologists to identify the direction from which forces on the rock originated, thus helping determine the causes of orogenies. Together, all of these facts about a metamorphic rock can tell us a good deal about its history.

Relative Dating

If we have discovered some events that have occurred in the past, the next thing to do is put them in order: what happened first, what second, and so on. This is called determining relative ages, or *relative dating* for short, because we are putting the events in sequence relative to, or compared to, each other. It doesn't tell us exactly *when* they occurred, only *in what order.*

Several basic principles help us with relative dating. The first we used already in Chapter 2. It's called the principle of superposition, and simply says that younger beds are on top of older beds unless the beds are overturned (which means upside down, a situation that occurs only rarely). This seems like a straightforward and logical principle today, but it wasn't so clear when everyone felt that all rock layers were the same age. A second important principle is that of *original horizontality*. It states that sedimentary rocks are deposited in a horizontal position; if we find them tilted, something has moved them since they formed. The original beds may not have been exactly horizontal, but very close; and that means drawings like Figures 4-1 and 4-2 actually exaggerate the tilt of the beds in order to make the drawings more clear. The Coastal Plain of Maryland is a good example of recently deposited beds which are still nearly horizontal; obviously the Valley and Ridge layers have moved since they formed.

A third basic principle is that of *original lateral continuity*. This is a concise way of saying that most sedimentary layers spread out over an area and are originally continuous, that is, unbroken by gaps or holes in the middle. If there's a gap in a layer, it's usually because part eroded away after the whole was formed. An application of this principle is shown in Figure 3-4, in the form of the dotted lines of the anticlines above the ground. These parts of the bed are gone now, but we know they once were there.

In drawings such as Figure 3-4 or 3-7, the dotted lines show a process called *correlation*. Correlation is the act of connecting beds which are physically separated, in order to show that they go together. Sometimes correlation is simply finding the same rocks in a new place and realizing they match up with another location—this is a relatively easy type of correlation. At other times, the rocks may be different, but we correlate them because they formed at the same time. An example of this would be the correlation of the rocks in Figure 4-2A or B: different types, but all deposited at the same time. This latter correlation would be hard to do if you couldn't see the connections between rock types—who would think that two different rock types are the same bed? This is one kind of problem that geologist detectives must overcome to determine the earth's history.

Of course, fossils can be a big help in relative dating. Fossils are the remains or traces of plants or animals preserved in rocks. There are many types of fossils, and a thorough discussion of them is beyond the scope of this book, but some of the references in Appendix G can give you more detail. For our purposes, however, it is important to note that the plants and animals of the world have evolved or changed over time, as indicated by their fossilized remains. Once we know what fossils existed at what time, when we find a fossil in a rock layer we are examining for the first time, we can use the fossil to date the rock. We can then also correlate it with other rocks of the same age, which should contain similar fossils. Note that the sequence of fossils by itself provides only a relative date; that is, it tells where the fossil and its rock layer fall in the succession of events, but doesn't tell us yet how old the rock is in years.

Another fact of relative dating is the *unconformity*, a place in a group of rock layers where part of the record of events is missing. It's as though you had a stack of newspapers that were in order by date from the bottom up, but someone had removed a few issues in the middle. For rocks, a missing section is usually caused by the area having been eroded, as shown in Figure 4-3. Thus, either no sediments were deposited to leave a record, or sediments which were deposited eroded away. Some unconformities represent very large gaps in the record. For example, horizontal sedimentary rocks on top of metamorphic ones mean the lower rocks were deposited, buried deeply, and metamorphosed; the rocks above them eroded away; then new rocks were laid on top of them. This sort of sequence could mean virtually a whole mountain (above the metamorphic rock) was eroded down to sea level, but no record is left of that long time period. Though the unconformity leaves the geologist with little specific information about a time period, it does tell him or her that erosion was occurring, and so the area was probably dry land instead of under water.

Finally, a principle called *crosscutting relationships* is also used for determining the relative age of a rock. This principle states that an intrusion or a fault must be younger than the rock in which it occurs. Obviously, a rock must already exist if a fault or intrusion is to occur in it. This tells us, for example, that the dikes in the Piedmont must have formed later than the rocks in which we find them; thus the time of the dikes' intrusion must be placed as an event separate from the formation of the surrounding rocks.

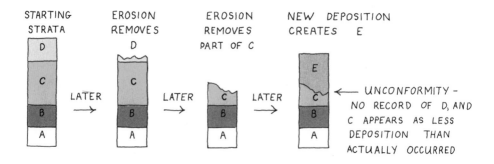

Fig. 4-3. Cross section showing the development of an unconformity.

Geologic Time

When geologists first began to put rocks in order by age, they assigned names to certain vertical sections of strata. Later, they figured out the age order of these named sections. (These names and the order of the geologic periods are listed in Appendix D.) However, they still had no firm idea of how many years ago each period started or ended, or how old the earth was as a whole. Determination of these values had to wait until radioactivity was discovered in the late 1890s.

The significance of radioactivity to geologic dating is as follows: Certain elements in the earth are unstable, and spontaneously change to another element (or a new version of the original element), giving off energy when the change occurs. The starting element is called the parent, and the ending material is called the daughter. The energy given off is the radioactivity, and this energy is the source of heat in the earth that keeps the interior hot. The change from parent to daughter occurs at a constant rate, starting when the rock containing a radioactive element is formed. Therefore, the longer time it has been since the rock formed, the more daughter there will be, and the less parent element. We can analyze a rock and find how much there is of both daughter and parent, and then calculate how long it has been since the rock formed. This is called the *radiometric age* of the rock.

An analogy for this might be what would happen if a man tore a piece of paper in half once a minute, and placed the discarded half of each piece on a table. The piece of paper in his hand (the parent) would constantly decrease in size, while the pile of torn pieces (the daughters) would constantly increase. By counting the number of torn pieces on the table, you could count how many tears had occurred, and so how many minutes had passed since the man started tearing. Because the radioactivity model is more complicated than paper-tearing, some inaccuracies are possible in radiometric dating, but with care we can obtain reasonable dates for the formation of certain rocks using this method. Then we utilize relative dating methods to get approximate ages for other rocks which can't be dated by the radioactive method.

After dating rocks this way, geologists could add the *absolute dates*—the ages in years—to the periods and the entire geologic time scale, as shown in Appendix D. It is interesting, though not easy, to try to get a grasp on the immensity of time that we have discovered in the age of the earth: at least 4.6 billion years. Various analogies have been used to convey the enormity of this time. Consider, for example, the time that has passed since the pyramids of Egypt were built, about 5,000 years ago. This seems like a lot of time to us, especially when we think of the changes that have occurred in human culture since that time. However, 5,000 years in 4.6 billion is the same as only 34 seconds in 1 year. Using a distance analogy, 5,000 years in 4.6 billion is the same as 17 feet in 3,000 miles (the distance across the United States). In those contexts, clearly all of human history is a brief instant in the age of the earth.

Developing a geologic time sense can do strange things to people's perspective. Everyday matters like getting to work or doing our laundry seem not to matter much in a time span that has gone so long without humankind, and presumably will continue with or without us. However, each mountain range or living organism now vanished had its part in this immense time, and we have ours, too. Also, we begin to see how dynamic and filled with changes the earth is, if we take a long enough view.

Sea Level Changes

Sedimentary rocks are often the result of material having been deposited in the ocean. Looking at the nature of the sedimentary rocks in Maryland, we can deduce that many areas of the state were once under water. Of course, these rocks weren't at their current elevation at the time they formed, but it still seems hard to believe that nearly all of Maryland was covered by a shallow sea (at least) at one time or another. However, that's the case, which brings us to a discussion of sea level changes over time—changes that involve hundreds of feet of elevation, occurring over thousands of years or more.

Some long-term changes in sea level are caused by real changes in the amount of water in the ocean. One way this can occur is through the formation and melting of *glaciers*. Glaciers are large masses of ice—sometimes more than a mile thick—which build up on land through an accumulation of snow during hundreds or thousands of years. When the climate becomes cool and wet enough for glaciers to grow in size, more and more water is removed from the oceans and tied up in ice on the land; so, sea level falls. The opposite happens when the climate warms and increased meltwater causes a rise in sea level. These changes have happened during several periods of geologic time, affecting Maryland's geology. Even now, we are in a time of rising sea levels, in part due to melting glaciers, which we'll come back to in Chapter 6.

Another way the sea level apparently changes is caused by alterations in the height of the land; in this case, what really changes is where the shoreline is, not the real worldwide depth of the ocean. This may happen in conjunction with the actual rising and falling of a continent. We don't usually think of the land as moving up and down, but sometimes it does (slowly!). (We'll see more such movement of the land in our discussion on plate tectonics later in the chapter.) If the land moves down, the shore will move in as the sea covers the land; if the land moves up, it will push the shore out. Another way to move the shoreline is by deposition and erosion. If there is a lot of deposition along a coast, the shoreline will fill in and move farther out, as though sea level were dropping. If erosion occurs at the shore, the sea will cover areas that used to be land.

Sometimes we can tell which one of the above situations caused a particular change in the shoreline position at a given time in geologic history; at other times we're not sure. In any case, the sea moving over the land and covering it is called a *transgression*. The opposite, the sea moving back and exposing more land, is called a *regression*. Each of these movements leaves a distinctive record in the rocks, and understanding them is important to deciphering the geologic history of Maryland.

The rock records created by transgressions and regressions are shown in Figure 4-4; we'll look first at how the transgression sequence proceeds, to understand the process and its results. First, layer 1 formed in a way similar to that of the layer in Figure 4-2B, with different sediments in

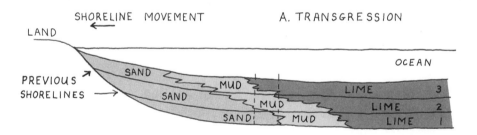

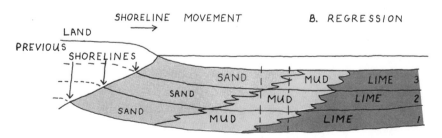

Fig. 4-4. Cross sections showing deposition patterns during changes in sea level.

different depths of water. Then the shoreline moved inland; so, when layer 2 formed, each sediment type was slightly closer to the land mass. Then the shoreline moved in again; thus, in layer 3, each sediment is even closer to land. Of course, the movement of the shoreline isn't in steps, but continuous. The important point to notice is what happens to the sediment type if we stay in one place as a transgression occurs. A good spot to observe is the area in the center of Figure 4-4A, between the dashed lines. That area is mostly sand on the bottom, then mud, and mostly lime on top. As we go up through the layers, we find sediments that are typical of deeper and deeper water. That makes sense, since if we stood in one place near the shore (for a thousand years!), we would expect the water to get deeper around us during a transgression. The principle that we learn here is "deep water sediments overlie shallow water sediments in a transgression."

A regression works the opposite way (see Figure 4-4B). Layers 1, 2, and 3 are deposited in that order, and each sediment type settles farther out as the shoreline moves seaward. Then if we look at our dashed-line area in the middle, we see the sediments changing from lime a the bottom, to mud, to sand at the top. Therefore, the rule is "shallow water sediments overlie deep water sediments in a regression." That sounds right, since the water should be getting shallower if we stayed in one place as the sea moved off the land. With this rule and that of transgression, we have a way to decipher Maryland's rocks, which will tell us when the sea got deeper or shallower over the state in the past.

Plate Tectonics

Some ideas in science don't agree very well with our everyday experience about the world around us, but are true nevertheless. We're not aware that the earth is round, for example, but we know that in reality it is a sphere. Similarly, we don't feel that the continents we live on could be moving across the earth, but they are, although very slowly. The first fairly complete statements of the evidence that continents move were made by Alfred Wegener in 1912, and he called this idea *continental drift*. The concept was not well accepted by scientists at that time, but new facts led to a reformulation of the idea in the 1960s. The new version was the result of many scientists' contributions, and is called *plate tectonics*. The word "tectonics" is a geologic term which generally refers to movement of rock, which in this case is the plates (described below), and the resulting structures. This theory plays an important role in describing and understanding the geologic history of Maryland.

Before examining the details of this theory, it may help to explain what the word *theory* means to geologists and other scientists. To most people, a theory is merely a guess; but in science, it is much more than that. A scientific theory is a concept that draws together and explains many verified facts, and is effective at predicting new ones. As long as a particular theory agrees with all facts, it remains valid; if new facts are learned that don't agree with the theory, then it is modified or a new theory is developed. Thus, though a theory may change over time, it still is our best explanation of the subject and is accepted as correct by nearly everyone. Though it seems incredible that continents can move, the plate tectonics theory explains a large number of otherwise inexplicable facts, and so is accepted by virtually all geologists today. Its power to draw a unified picture of what makes geologic events happen, and to account for geologic realities, has revolutionized geology in the past few decades.

Plates and Their Boundaries

The theory of plate tectonics states that the outer layer of the earth, called the lithosphere, is not one continuous piece, but instead is made up of separate plates, which can move around on the earth. (The major plates and their current direction of motion are shown in Figure 4-5.) These plates are

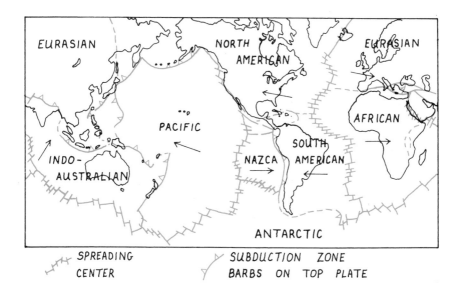

Fig. 4-5. *Major tectonic plates of the earth and their current direction of motion. Adapted from a map by the U.S. Geological Survey.*

free to move because they float on a layer of rock which is hot enough that it can flow or move slowly, like taffy or thick mud. The idea that the continents are floating is called *isostasy.* Supported by the underlying material, the plates can move slowly across the earth in response to forces pushing on them. They also may move up or down somewhat, in response to changes in their weight or in other pressures on them.

The top layer of the plates is called the crust, and the crust under the oceans is different from the crust that forms the continents. The crust under the oceans is made of basalt, and is relatively thin and dense. This density depresses the plate under the oceanic crust, so the plate floats low, like a solid, heavy log. This depression creates the ocean basins, where water can collect. Meanwhile, the continental crust is mostly made of the felsic minerals found in granite, and so it is less dense, but thicker, than the basaltic ocean crust. Thus, the part of a plate carrying a continent floats higher than areas of oceanic crust and so stands up above sea level. Most plates carry both oceanic and continental crust, but both parts of the plate can move together as a unit.

The plates are rigid and together they cover the entire surface of the earth. If new plate appears in one area, existing plate must disappear somewhere else. The new plate that appears is in the form of basaltic magma that rises from deep in the earth at places called *spreading centers,* shown in the center of Figure 4-6. These are actually breaks, or rifts, between plates where magma flows out as lava, pushing the plates away from the rift as new plate is formed; reflecting this movement, these rifts are also called *divergent boundaries.* Most of these breaks are under the oceans, so we are rarely aware of these underwater eruptions. One major rift runs in a curving, north-to-south path through the middle of the Atlantic (separating the

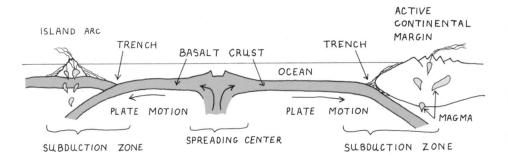

Fig. 4-6. Cross section of a spreading center and a subduction zone.

American, African, and Eurasian plates in Figure 4-5), and this type location is why these features are sometimes called *mid-ocean rifts.*

On the far side of the plate away from the rift, the plate is being pushed against adjacent ones, creating a *convergent boundary.* Here, one of the plates must get pushed down into the earth to make room for the spreading from the other side. The area where a plate moves down into the earth is called a *subduction zone,* and structures there vary somewhat depending on the type of crust found at the plate boundary. Subduction zones complete the appearance-and-disappearance cycle of the plates, which is the fundamental pattern of motion that keeps the earth geologically active.

Several types of convergent boundaries exist, depending on what types of crust are coming together. If both plates are carrying oceanic crust at the boundary, then either plate at the boundary could be pushed down into the earth. If one plate is carrying ocean and the other is carrying a continent, the continent will be too low in density to be pushed down, just as a cork is hard to sink even in a whirlpool. Instead, at such a boundary, the plate that carries the oceanic crust will be pushed down into the subduction zone, and the continent will remain on the surface.

The subduction zones in the above two cases create distinctive landforms. In both, where the plate moves down into the earth, it creates a deep part of the ocean, called a *trench* because it has a long, narrow shape running along the plate border. These trenches are the deepest parts of the oceans. The basalt crust and sea floor sediments on the subducted plate are heated when they move into the earth, and melt to create magma. This magma rises to make granitic intrusions and volcanoes composed of rhyolite lava and related rocks. At a boundary where two pieces of oceanic crust converge, the volcanoes created by the subduction may rise high enough to become islands. The string of volcanic islands often makes a gently curving line, giving this type of subduction zone the name *island arc,* as shown on the left side of Figure 4-6. The western Aleutian Islands and many of the islands in the western Pacific are of this type.

If the subduction zone occurs where oceanic crust is being pushed under a continent, a trench will still be created, just offshore from the continent. In addition, the pressure of the converging plates will crinkle the edge of the continent, making a mountain range there. Any mountain-building event is called an *orogeny* by geologists, but the causes of orogenies were not understood for a long time. Now we know that a convergent boundary between plates is the place where most orogenies occur. Mixed in with the mountains at a continent-ocean subduction zone will be volcanoes, formed in the same way as the island arc volcanoes, but this time on the continent.

This type of continent edge is called an *active continental margin,* and is shown on the right side of Figure 4-6. Two current examples are the Andes Mountains, created by subduction of the Nazca plate under the South American continent, and the Cascade Range in the northwestern United States. Sometimes the pressure on the edge of the continent causes an area farther inland from the mountains to buckle *down*, and an inland sea forms there. Sediments washing off the mountain range accumulate in this sea, and may form very thick deposits. We will see this type of accumulation when we examine the geologic history of Maryland.

The third type of convergent boundary occurs when two continent-carrying plates are pushed together. Since both crusts are similar in density, one can't be pushed back into the earth like a plate that's carrying basalt, so the two continents literally collide. Though the speed of the crash isn't like that of two cars, the same kind of thing happens: the edges get crumpled, and the two land masses are welded together by the force, making one large mountain range, as shown in Figure 4-7.

In this type of orogeny, the greatest pressure occurs at the center of the new mountain range. There we may see intensely folded rocks, sometimes even bent back on themselves like paper folded in half. The pressure also causes a large amount of metamorphism in the rocks near the core of the range, and may melt some rocks to produce intrusions as well. And, as would happen when two bulldozers collide head-on and so trap dirt between them, the continents may squeeze some basaltic ocean crust into the core of the range. The pressure also affects areas distant from the actual edge of

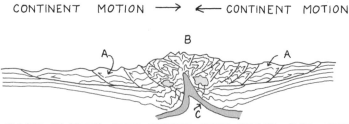

CONTINENT   MOTION ⟶   ⟵ CONTINENT   MOTION

A: BLOCKS OF FOLDED ROCKS, THRUST ALONG FAULTS OVER OTHER ROCKS.
B: PREVIOUS EDGE OF CONTINENTS, NOW CENTER OF MOUNTAIN
   RANGE; INTENSELY FOLDED, METAMORPHIC ROCKS,
   AND SOME IGNEOUS INTRUSIONS.
C: OLD OCEANIC CRUST CAUGHT IN COLLISION.

Fig. 4-7. Cross section of a mountain range formed by the collision of continents.

the previously separate continents. Rock layers can be pushed back over other layers along deep thrust faults, like the blocks labelled A in Figure 4-7. Geologists sometimes call this faulting "thin-skinned" tectonics because the pieces being pushed over each other are thinner than complete crust. Actual blocks like these in the Appalachians are thousands of feet thick and have been pushed more than 100 miles over underlying rocks by continental collisions. This motion folds the layers of the top block and probably does some folding of the layers below the fault as well. Overall, the folds become less intense the farther they are from the center of the mountain range, in the same way folds decrease if we travel in Maryland from the Piedmont to the Allegheny Plateau.

Of course, not all pieces of continental crust are large, so some continent-continent convergent boundaries will involve a small landmass hitting a much larger one. The small landmass could even be an old island arc which is no longer active. The smaller landmass will be folded and squeezed in the collision, and welded onto the larger continent. Further collisions with more landmass fragments may result in several blocks of land, each bounded by faults, welded onto the edge of the large continent. Geologists call each of these blocks a *terrane*. Because a terrane is a piece of a different landmass than the larger one in which it is found, its rocks may be quite distinct from the adjacent rocks outside the terrane. Terranes which have formed in this way seem to make up parts of Maryland's Piedmont.

As continents push against or over ocean plates, an environment for particular types of rocks is created. Imagine the mixture of broken wood, brick, concrete, and other debris that collects in front of a bulldozer pushing over the site of a demolished building. Where continental and oceanic crusts meet underwater, the movement will fold the sediments into a similar jumble of mixed rock materials. These materials will contain broken, angular fragments of a variety of rocks, mixing all sizes regardless of whether the sediment came from the ocean bottom, or slid down from the edge of the continent above. This type of rock body is given the French word for mixture: *mélange*. A rock type often found in a mélange area is *diamictite*, which is composed mostly of relatively fine grained sediment, but also contains larger chunks of various rock types. This rock is somewhat like conglomerate or breccia, but naming a rock diamictite often implies it has an origin in tectonic movements, not in running water, volcanoes, or other sources. Diamictite also may be somewhat metamorphosed by the pressures that created it. Actually, geologists are not exactly sure how diamictite forms— more study is needed. Mélanges and diamictites are not common, but do exist in Maryland, and add evidence of an active geologic history.

That covers the main points concerning convergent boundaries, but we need to look back at a few more points about divergent plate boundaries: Sometimes a spreading center forms under an area of continental crust. If it does, it will tear the plate and its cover of continent apart, turning it into two continents on separate plates. Between the two, the land will sink down, or subside. There may be some basalt volcanoes on the surface of this new rift, and basalt-composition intrusions may occur. Sediments will also accumulate in the rift. The lowest place on the earth—the Dead Sea—is an example of this type of area, as is the Rift Valley of eastern Africa. Eventually, these areas open to the sea and become flooded. As the rift continues to widen, the body of water becomes a new ocean between the two continents, and the rift edges become shorelines. The spreading center that caused the rifting will now be a mid-ocean rift.

As these rifted continents move apart, both the continental crust and the oceanic crust will be moving in the same direction, that is, away from the rift. This situation is shown in Figure 4-8. Because there is no relative motion between the crust types, sediments that wash off the continent and into the ocean can settle down in undisturbed, nearly horizontal layers. Offshore, it may be a good location for carbonate sediments to accumulate and make a carbonate bank. Overall, this is called a *passive continental margin* due to its lack of tectonic activity. A look at Figure 4-5 shows that Continental Shelf of Maryland is in the correct position to be a passive margin, which explains why this broad shelf of sediments exists. In contrast, because there *is* relative motion between the ocean and continental crust along much of the Pacific side of North America, there are fewer continental shelf deposits along the West Coast of the United States.

What force drives all of these plate movements? Whatever it is, it is so deep in the earth that scientists can only make educated guesses about it. Their best answer is that heat (probably from natural radioactive energy release) softens and moves the rock deep in the earth below the plates. This

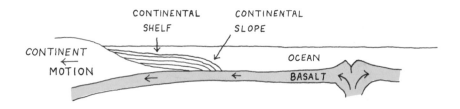

Fig. 4-8. Cross section of a trailing edge of a continent.

creates motion like convection cells similar to, but much slower than, the circulation of vegetables in a boiling pot of soup. This motion drags the plates around, like slabs of ice moved by the current in a river, pulling them apart or pushing them together. It also supplies the magma which rises at the rifts created when the plates are pulled apart. Apparently these motions began shortly after the earth formed, and will continue as long as radiation in the earth keeps supplying heat, which should be for a long time. The speed of the plates varies but is about 1.5 inches (4 cm) a year on average; that's about as fast as your fingernails grow.

A Sequence of Plate Tectonics Events

Plate tectonics is useful not only for understanding individual features of the earth, but also for showing how interactions between plates cause changes over time. Specifically, we can use plate tectonics to construct a sequence of events that explain mountain ranges like the Appalachians that we see in Maryland. These events are shown, generalized and idealized, in Figure 4-9, and described below. To make this pattern easy to apply, the events are described as they occurred in Maryland, but the description holds true for other places around the world as well.

Imagine that a long time ago there was a large landmass under which a spreading center formed. This stretched the continent and cracked it, allowing basalt up through the cracks as blocks of the continent began to subside (Figure 4-9A). Eventually, this rift opened enough to allow an ocean to grow between the now-separated continents (Figure 4-9B); something like this may have happened to North America (*at left*) and an unknown continent (*at right*). As the two continents separated, sediments accumulated as a continental shelf at their passive margins (Figure 4-9C).

Then a change of direction occurred, and the continents began to move together again, as shown in Figure 4-9D. A slight motion might have buckled the passive margin sediments up into a small mountain range at the coast, as on the left of Figure 4-9D. More movement caused real subduction, creating an active continental margin, shown on the right side of Figure 4-9D. The continents continued to move closer together; eventually they collided, and built a mountain range. This also welded the passive margin sediments onto the continental edge as folded rocks, forming one large continent again. The collision created the same structure we saw in Figure 4-7: metamorphic rocks in the middle and folded and faulted rocks on both sides.

A large continent traps heat rising from deeper in the earth, and this pressure created a new rift near the place where the two continents came

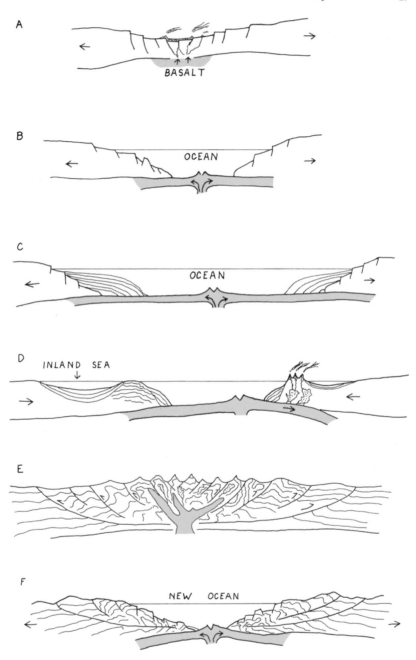

Fig. 4-9. Cross sections of a sequence of plate motions that make mountain ranges near coastlines.

together. This would have broken the mountain range approximately in half, creating a mirror image on the two landmasses. (In fact, parts of the Appalachians of North America are mirrored, generally, by the Atlas Mountains of northern Africa and the Scottish Highlands of Great Britain, indicating these areas fall into this sequence well.) As the rift activity continued, the ocean between the land masses widened. The mountains near the edge of the continents eroded and deposited their sediments offshore on the new passive margin, returning us once more to Figure 4-9C.

Though, again, the events given above are simplified and generalized, notice that we have built up the general geologic structure of Maryland by this sequence. We have folds to the west (Allegheny Plateau and Valley and Ridge provinces), metamorphic rocks closer to the ocean where the old core of the mountains was (Blue Ridge and Piedmont), and sediments accumulating along the shore (Coastal Plain). The opening, closing, then reopening of the Atlantic Ocean basin like elevator doors welded the sediments that were in the ocean basin onto the continents. This made the continents each a little larger than they were before, a situation that agrees with the fact that the base rocks of the continents seem to be made of old, eroded mountain ranges, with older ones at the center of the continent and younger ones on the edges. So it seems like the continents of the earth have grown by accretion, adding a little bit on the edges each time they moved around and collided with one another during the last few billion years. This is the way a snow pile grows as a snowplow pushes another plowful onto the side of the heap. If we speed up the time scale, we can imagine continents bouncing off each other like billiard balls on a table, adding a bit more material each time they collide.

All of the above is a neat, simple explanation—a bit too simple. The emphasis here has been on the overall nature of the continent-motion cycle and some of its possibilities, but we will see in Chapter 5 that the actual occurrences in Maryland have been more complicated. Also, we don't know all there is to know about the rocks of Maryland and North America, and new facts may change our interpretation of the events. This may involve only details, or it might require major revisions. However, the above general sequence, and the details that appear in the next chapter, are a sensible sequence we can start with, recognizing that it's subject to change.

Now we can look in some detail at the changing conditions in Maryland during the last billion years that have created the individual rock formations that we see today.

Chapter Five

The Geologic History of Maryland

The story of how the rocks in Maryland came to be the way they are is a long one. To make it more manageable, it can be divided in sections according to age. Though any division of a continuous sequence of events is arbitrary, we'll start with the Precambrian, then go over the events in the Paleozoic up until the mountain-building event called the Taconic orogeny. After that, we will use the cycles of mountains forming and eroding as our segments, dividing them into separate time frames each time new mountains form. This discussion will cover how the land and environments changed over time—mountains, lowlands, beaches, or sea—with mention of the major rock units that formed in those environments. Of course, it is the rocks themselves that tell us what conditions prevailed at the time they were formed.

One major fact that applies to all of Maryland at the start of this story is that North America as a whole was closer to the equator in the past, as well as turned in a different direction. For example, the continent probably was in such a position that the equator ran roughly from North Dakota to Texas at one time, and in other positions at other times. This means Maryland may even have been south of the equator during some time in the past. The evidence for this is the types of plant and animal fossils found in Maryland, and other pieces of physical (nonbiological) evidence. Therefore, it was warmer in Maryland overall for the first few hundred million years covered in this story. This means today's "western Maryland" was actually "northern Maryland" when the continent was oriented differently. In this discussion, however, let's ignore this last fact, and use directions as they are in the state today. From here on, though we concentrate on how continental motions and collisions affected Maryland, these events moved North America as a whole to its current position on the earth.

Also, as mentioned earlier in this book, the rocks on the surface of the Appalachian provinces probably formed significantly farther east of where we find them today, and were later pushed to their current location by orogenies. However, to keep things simpler, we'll refer to their environments of deposition as though they occurred in their current location (with some exceptions).

This story begins with the earliest rocks, and moves forward in time. The geologic column, that is, a list of the rock formations in age order, is in Appendix E, which also includes a basic summary of the history. In some cases, geologists are not sure of the time of events because the evidence isn't definite. Note also that, although there are times given for the orogenies, they often occurred in pulses, with different events happening in varying places at different times. Because of this, we must be careful not to apply the history mentioned here to Appalachian areas distant from Maryland. Specifically, geologists have long noticed that the way the rocks in New England have been affected by orogenies is different compared to the central (where Maryland is) and southern Appalachians, so events were probably different in these areas.

There is a phrase in geology that says, in effect, "The geology of an area doesn't change much, but the interpretation can change a lot." This means that geologists can re-examine an area already studied, and interpret it quite differently than was done before. In this process, formation names can change, formations' age order can be rearranged, and formation boundaries on geologic maps can be redrawn. Many areas of Maryland have had two or more careful studies done on them, with each study taking years of work by several scientists, yet unanswered questions remain. The information in Appendix E and the geologic history are based on the current interpretation, which may change in the future. This is a normal part of science, and we can hope our knowledge improves as time goes along.

By way of warning and encouragement, I should mention that the beginning of this story is the most complicated section. This is partly because more things happened in more varying places near the beginning of the sequence than later. Also, the fact that the start was so long ago means that much of the evidence of what happened has been altered, and so is hard to interpret. The result is that we sometimes have more than one possible way the events may have occurred. However, the complexity is fascinating, and things do become more straightforward later.

Maryland's Early Geologic History

The geologic history of Maryland starts far back in the Precambrian era, more than a billion years ago. At this time, North America was smaller than it is now, and the eastern shoreline was farther west than today, probably at least in the middle of Maryland. Sediments, perhaps partly of volcanic origin, were deposited offshore from this coast, became deeply buried, and metamorphosed into the Baltimore Gneiss, Maryland's oldest rock. Similar sediments may have been the starting material for the Middletown Valley Gneiss, or it may have formed from an intrusion of igneous rocks at this time. The metamorphism or intrusion of these rocks occurred about 1.1 billion years ago, in the Grenville orogeny; this event gives us our first firm date in the story. This orogeny may have been caused by a collision of North America with another continent. Such a collision would mean Maryland was probably mountainous, and in the interior of a large landmass in the later Precambrian.

A long period of erosion then followed, leaving no record of events, and wearing away the mountains made in the Grenville orogeny. Eventually, enough materials were removed to expose the rocks at the core of these mountains, the Baltimore and Middletown Valley gneisses. Thus, the top surfaces of these rocks represent a major unconformity, with much time missing.

In the late Precambrian, volcanic action began in what is now the Blue Ridge and Middletown Valley area, and resulted in the basalts, tuffs, and rhyolites of the Swift Run, Catoctin, and Sams Creek formations. This probably indicates rifting of the continental landmass at the time, and it appears that the rifting occurred in more than one place, breaking the continent into several pieces by Cambrian time. Since the rifting process took a good while, the volcanic deposits were exposed to enough erosion to form weathering surfaces.

Some of these rifted pieces are shown in Figure 5-1A, labelled BR for Blue Ridge and western Piedmont, BG for Baltimore Gneiss (central Piedmont), and EP for eastern Piedmont. These names reflect the approximate relative locations of the blocks today, though geologists are still collecting evidence to decide how many separate pieces there were, where they separated, and how they have moved over time. As rifting continued, the basins separating these blocks widened enough to allow the ocean in, forming a body of water that geologists call the Iapetus Ocean. Since this ocean off the east side of North America was in the location of the current Atlantic Ocean, it is named for Iapetus, who was the father of Atlantis in mythical stories. We will apply the name Iapetus to all of the ocean east of North America, including both sides of the EP block and continuing east until we reach another continent.

As time progressed, the edge of the continent may have subsided, so the shoreline would have moved inland to somewhere in the middle of the

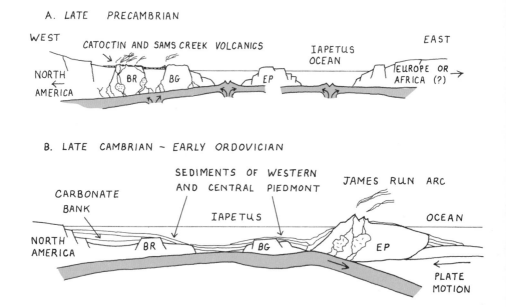

Fig. 5-1. *Generalized cross sections of possible steps in the geologic history of Maryland for late Precambrian to early Ordovician time. Note that this and all later cross sections in this chapter are vertically exaggerated to show detail.*

state. Farther seaward, the continental blocks also subsided, and therefore were covered by water and began to accumulate sediments. The rifts among the blocks became less active, so each block remained more fixed compared to the continent, at least for a time. But the separation of the blocks meant they had different environments, so a variety of rocks were created as deposits formed off the edge of the continent, on top of submerged blocks, and in the basins between blocks.

Looking at the Blue Ridge block area, the first sediments were sands, probably laid down by rivers or on beaches which later became the Weverton and Sugarloaf Mountain Quartzites. These were deposited in late Precambrian or early Cambrian time. Then the Harpers, Urbana, and Ijamsville formations were deposited in the western Piedmont area; as it became covered with deeper water they formed the continental shelf. Meanwhile, during this time the offshore Baltimore Gneiss block probably was subsiding, so eventually it also had a shallow water cover where the sands and thin muds of the Setters Formation were deposited.

During early Cambrian time in the western Piedmont, the water probably became shallower, so wave action could transport and sort the sediments to create the sandy Antietam Formation. Farther east and offshore, mud and silt probably in deeper water made the Araby and Cash Smith formations. Even farther out, the water depth remained shallow enough on the Baltimore Gneiss block to make the Cockeysville Marble.

As the nearshore shelf stabilized, a carbonate bank developed (shown in Figure 5-1B) that included much of Maryland west of the western Piedmont. The bank existed continuously into Ordovician time, and built up a thick accumulation—more than a mile total—of limestones and dolomites. This bank includes all of the formations between the Tomstown Dolomite and the Chambersburg formations in the Valley and Ridge, with the exception of the Waynesboro, which was a temporary return to clastic sediments. As the carbonates accumulated, the land subsided, but the surface was always close to sea level; therefore, small changes in water level alternately caused transgressions and regressions. The bank as a whole shows many cycles of water depth change, though it was never under deep water. As this bank built seaward, it extended the conditions where limestones could form, so the Frederick and Grove limestones accumulated in the area of the western Piedmont.

All of the Cambrian schists in the central Piedmont—Marburg, Gillis, Prettyboy, Pleasant Grove, Loch Raven, and Oella—probably formed in the area east and offshore from the carbonate bank at the edge of the continent. Their sediments were only slightly different in varying areas of the basin, so the rocks are generally similar with slight variations. The Marburg was close enough to the carbonate bank to include carbonate deposition at times, resulting in the Silver Run Limestone Member. East of the Marburg, the muddy sediments of the other schists accumulated in fairly deep water, in a basin between the continent and the Baltimore Gneiss block. Even this block was probably still subsiding, however, so similar muddy sediments were deposited on the block to become the Loch Raven and Oella formations.

Meanwhile, the spreading that had pushed the eastern Piedmont block away from the continent changed to subduction, as shown in Figure 5-1B. This subduction may have started as early as the beginning of the Cambrian period, though it took time for an island arc to grow. In fact, the eastern Piedmont block may have been small, or even nonexistent, so most of the block may have formed from the subduction activity mentioned above. This subduction was the source of volcanic materials and intrusions which made up the James Run and Port Deposit formations in late Cambrian or early Ordovician time. These deposits are substantial, being roughly a mile thick. The highlands and deposits made by this subduction are another orogeny (sometimes called the Avalonian orogeny) which contributed to the formation of the Appalachians, but these formations were not an actual part of Maryland until they were pushed onto North America in the Taconic Orogeny.

Also as part of this subduction there were intrusions of mafic magma, which became what is now the Aberdeen Gabbro and the Baltimore Mafic Complex. The magma probably was intruded beneath the land which was becoming the island arc, though it is also possible that there were intrusions closer to North America, into the Prettyboy and Pleasant Grove formations. The reason that it is hard to determine the intrusion location is that these mafic units were to be moved around significantly by orogenies yet to come in our story. On the map in Figure 3-10, these rocks generally are the ultramafics found in western Baltimore and northeast of the city.

Taconic Orogeny and Erosion

By Ordovician time, pressure from the east began to push the block of the James Run island arc shown in Figure 5-1B against the other blocks to its west. Continued westward motion of this arc created unstable and active conditions on its leading (west) side, causing landslides which occurred both above and below the water. Thus, many parts of the Liberty Mélange rocks contain blocks or clasts of other rocks mixed in a jumble of sediments. The Morgan Run Formation, itself made of underwater landslide deposits, was lifted out of the sea as the island arc collided with the Baltimore Gneiss block, and partly eroded in more landslides, which were deposited as part of the Sykesville Formation.

Also mixed into the Sykesville were pieces of oceanic crust which originally were between the blocks in Figure 5-1B; this crust appears as ultramafic rocks in the mélange. The ultramafics may also include material from the mantle, which is the next layer of the earth below the crust. The largest piece of this material we now see on the surface is the serpentinite of Soldiers Delight. Many of the small bodies of ultramafic rock shown in and around the Sykesville in Figure 3-10 are probably other pieces of oceanic crust. All of the ultramafics in the Piedmont give us some insight into the nature of the materials which underlie the oceans and occur above subduction zones, for they are much easier to study now that they are on the surface instead of in the inaccessible locations where they formed.

Some parts of the Sykesville and Morgan Run formations are schists that look similar to the Loch Raven and Oella formations. These rocks plus the Prettyboy and Pleasant Grove all used to be considered together as the Wissahickon Group, but geologists now believe the locations where the various rocks formed are quite different, so the group name is no longer applied. Future mapping and reclassifications may make changes to the areas shown for these formations in Figure 3-10.

By late Ordovician time, the westward motion of the eastern Piedmont island arc had pushed the Baltimore Gneiss and Blue Ridge blocks up against the continent, and up out of the sea. This pressure also buckled the seaward edge of the carbonate bank down, so the water just east of the continent became deeper than before. Into this basin, sediments flowed westward, off the new land to the east. The first sediments began to mix with the limestones to make the dark, muddy limestones of the Chambersburg. Then the clastic sediments became dominant and made the black Martinsburg Shale. There are volcanic ash layers in both of these formations, indicating there were some volcanoes as part of the new highlands. The pressure from the east was very strong at this time, and began to break apart and fold the rock layers in the Blue Ridge and east of it. This created

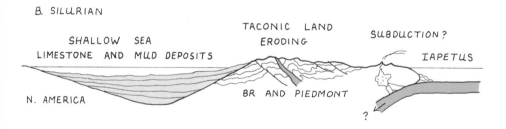

Fig. 5-2. Generalized cross sections of possible steps in the geologic history of Maryland for the Taconic Orogeny and erosion that followed.

true mountains in the area (shown in Figure 5-2A), an event called the Taconic orogeny. Actually, the pressure had been building since Cambrian time, pushing the continental blocks closer over an extended period. In the Taconic orogeny, they all came together.

High mountains mean increased erosion, and large amounts of sediment began to flow westward and fill the basin. At this time, the basin itself was probably being lifted from the pressures of the orogeny, while sea level may have dropped due to glaciation on other continents. These factors combined to make the deposition become nonmarine: the Martinsburg Shales change into the alluvial fans and deltas of the Juniata Formation. The Juniata is red, as are many alluvial deposits on land, an important indicator that western Maryland was now dry land. At this time, there were plants and animals only in the sea and none (that make significant fossils) on land; that the Martinsburg contains fossils but the Juniata does not further suggests the latter formation is nonmarine. The Martinsburg and Juniata are the bottom of what we will see, as this story progresses, is a large stack of sediments eroded off the Taconic highlands. This stack of sediments is called the Queenston Clastic Wedge, which partly covers Maryland, Pennsylvania, New York, and New Jersey. These rocks are at the bottom of what is called the Appalachian Basin, which is the area west of the Appalachians where sediments began to collect at this point in our story, and continued to do so for some 200 million years.

The Piedmont itself is where the Taconic highlands were, so no new rocks were deposited there at this time; however, the pressures of the collision rearranged the existing rocks substantially. It's useful to look in some detail at these effects, because we will see similarities in later Maryland orogenies.

One effect was metamorphism: The igneous James Run and surrounding sedimentary deposits were changed to metamorphic rocks at this time, though they were also affected by later orogenies in order to reach the condition in which we see them today. The rocks around the Baltimore Gneiss were also metamorphosed, perhaps even before Taconic time, as the blocks were pushed together. Overall, the metamorphism of Piedmont rocks was significant, and probably occurred in various places at several times in the Ordovician period.

The heat and pressure of metamorphism made some rocks become softer and flexible, a condition which geologists call *plastic* or *mobile*. What happens then is like what happens to the filling when a jelly doughnut is squeezed: the filling—in this case the mobile rock—pushes out in any direction it can. In rock, the direction of motion usually isn't downward because the rock would encounter increased pressure there. Instead, it moves in a direction somewhere between up and horizontal. The rock in Maryland that shows this behavior best is the Baltimore Gneiss, which was mobilized and squeezed generally up and northwest in the Taconic orogeny.

As the mobilized gneiss pushed up, it was folded, and caused folding of rocks around it, especially the Glenarm Group (Setters, Cockeysville, Loch Raven, and Oella). This folding, coupled with faulting, was intense enough to create nappes, and they pushed west over, under, or through other rocks. The warping and twisting of these nappes in several directions at once gives us the uneven surfaces now exposed by erosion as domes (though we should recall from Chapter 3 that these are not all truly domes). These forces also caused the largest amount of metamorphism to occur near the center of the domes. On a larger scale, the folding created a broad upwarp area, running from central Baltimore County southwest toward Washington, known as the Baltimore-Washington Anticlinorium. Slightly less intensive folding bowed up the anticlinorium in the Sugarloaf Mountain area and in the Blue Ridge.

Finally, we also see the effects of the compression from the east in a large number of thrust faults, which also moved rocks over other units to their west. Many of the rock contacts between formations in the eastern Piedmont are thrust faults which were then active. We can examine this by looking at the rocks that we find as we move northwest along the western side of the Susquehanna River, starting from the Chesapeake Bay (Figure 3-10 shows the rocks involved, though not all of them are separated). These rock units generally dip southeast and each unit sits on top of the unit to its northwest, with a thrust fault between them. The Aberdeen Gabbro is a sheet pushed over the James Run–Port Deposit, which is over the Conowingo. Continuing northwest, the Conowingo is over the Baltimore Complex, which is over the Sykesville, which in turn is on top of the Morgan Run. The Morgan Run overlies the Prettyboy, and the Prettyboy probably is thrust over the Gillis. The Gillis is part of the Linganore Nappe, which has pushed over the Marburg and units below it. Interestingly, the metamorphism of formations like the Cash Smith and Araby is primarily a result of shearing or tearing forces, probably caused by these thrust fault motions. Clearly, the Taconic orogeny squeezed many rocks together, piling them up on the edge of North America.

Much of the Piedmont we see today may have been buried under several of these thrust sheets that have now eroded away. For example, we find the Sykesville Formation both west and east of the Baltimore Gneiss areas. This suggests that the Sykesville (and probably other units as well) may have been a larger sheet that once covered the whole "domes" area, but now is eroded away in the middle to expose the rocks below. The nappes and thrust sheets can be considered to be terranes which have been pushed together to make the Piedmont, though the boundaries among separate terranes are difficult to place. The complex folding and stacking of these sheets, along with the orogenies and uneven erosion that have occurred since they were put in place, accounts for the complicated patterns we see in the Piedmont rocks today.

The last part of the Taconic orogeny left to describe is the formation of the granites. Heat and pressure of the now-thickened crust on the edge of North America melted rock in the central Piedmont, which made the intrusions of the Woodstock, Ellicott City, and Guilford granites in the late Ordovician or early Silurian period.

A sequence of events for the Taconic orogeny having been described, it is important to note again that this is only one of several possible sequences. As mentioned earlier, it is difficult to say exactly how many pieces the continent rifted into in late Precambrian time; therefore, their movement history could be different. Geologists are fairly sure that the James Run and surrounding rocks came from an island arc, but where the arc formed is debatable. It could have been an attachment to the eastern edge of the North American continent, or a separate block offshore, or hooked to the western edge of a continent which was east of North America; Figure 5-1B takes the middle choice as being the easiest to mentally modify if future evidence requires it. The age of the Glenarm Group has been debated for a long time, and is still not definite; some of the rocks of this group may have been thrust to their current position instead of forming on top of each other. It is possible that the Baltimore Gneiss block formed north of Maryland, or even as part of another continent, and was pushed to its current location during the Taconic orogeny. Overall, further study is needed to make our knowledge of the Taconic orogeny more definite; but until then the above story is at least generally accurate.

By early Silurian time, a new sea had begun to develop west of the Taconic mountains, as in Figure 5-2B. As these mountains eroded, the resulting sediments were carried west into this sea (and probably east as well, but we have no record in the rocks of this). The first deposit in the inland sea, the Tuscarora Sandstone, was at least in part a beach deposit formed as the shore transgressed east towards the mountains. This is one of the ridge-making layers we looked at in Chapter 2; it is interesting to see that a beach deposit of Silurian time helped to make the mountains of today. The next beds (Rose Hill, Keefer, and McKenzie) all show deeper water sediments in the west and shallower ones in the east. The basin filled with sediments that became the red beds of the Bloomsburg; but then shallow water returned, where the Wills Creek Formation was deposited. After that, the water was deep enough again to initiate a series of mainly limestones that continued into the Devonian period (Tonoloway Formation through the Helderberg Group).

The situation during the middle of this series of limestones is shown in a map view in Figure 5-3. Notice that this is the opposite of the present condition in Maryland: Today, the mountains are in the west and the sea is to the east; during Silurian time, western Maryland was underwater, while mountains were in the east, where the Piedmont and Coastal Plain provinces are today. The sediments that water carried west from these mountains and deposited in shallow water or on land as alluvial deposits make up the rock layers listed above. Meanwhile, there are no rocks from the late Silurian and into the Devonian in the Piedmont, because that was an area of erosion, not

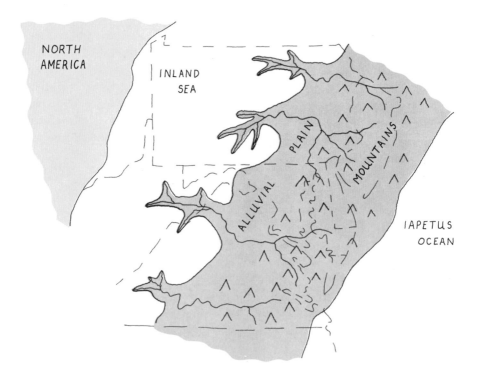

Fig. 5-3. General geography of the mid-Atlantic states at the end of Silurian time.

deposition. Figure 5-3 is only meant to give a general idea of the geography, as the precise location of shorelines changed constantly over time. Also, recall that the mountains and inland sea both may have been farther east, as later orogenies perhaps pushed all of these rocks west of where they formed.

Figure 5-2B also shows another environment east of the remnants of the Taconic higlands—a zone of subduction and volcanoes. Although there is little or no direct evidence in Maryland that there were volcanoes at this time, we do know the landmass(es) east of North America moved toward the west, so the ocean crust between them had to be disappearing somewhere. Perhaps the subduction was beneath the other landmass(es), perhaps not. In any case, this possible subduction, also shown in Figures 5-4 and 5-5A, reminds us that areas east of North America were moving toward the continent.

The limestones of the Helderberg Group show that not many sand and clay sediments were eroding into the inland sea by the early Devonian, so the Taconic mountains were nearly worn away. Thus, their part in Maryland's history ends, and a new cycle begins.

Acadian Orogeny and Erosion

As just mentioned, the shorelines shown in Figure 5-3 weren't stationary. As the water moved over Maryland, perhaps more than once, it deposited the quartz of the Shriver Chert and Oriskany Sandstone. This shows the water level overall had regressed from that needed for the previous limestones; in fact, there was a short period of erosion when there was little or no water cover over the state. This change in conditions signaled the start of a new orogeny early in the Devonian, called the Acadian orogeny.

The precise cause of the Acadian orogeny in Maryland is not clear. In New England at this time, there were intrusions, volcanoes, and folding/faulting/metamorphism, with possible collision of some continental fragments. Farther north, Europe was approaching northern North America, and pressures of collision were occurring there. To the south, from Georgia to Virginia, there was a significant amount of metamorphism, and perhaps collision with continental fragments. Thus, there was activity all around Maryland, but little evidence of any direct cause for mountain-building here. We are left to conclude that eastern Maryland was simply uplifted by pressure from adjoining areas. In this, Maryland was like the side panel of a car in a head-on collision—it still got crumpled even though it was not in actual contact with the colliding object.

Another way to see why the cause of the Acadian orogeny is uncertain in Maryland is to look at the physiographic province map, Figure 1-2. Notice that the Piedmont in North Carolina is fairly wide west-to-east, as is the New England province across Connecticut (these two areas are structurally similar at a general level). Compared to these areas, the Piedmont in Maryland is much narrower. It is in these wide areas in the southern Piedmont and in New England that we find direct evidence of the Acadian orogeny, but this region is missing in Maryland. Maybe there really were pieces of continent that collided with Maryland at this time. If so, the now-absent part of the Piedmont in Maryland might be under the Coastal Plain and Continental Shelf. It also has been suggested that this part of Maryland was pushed south by a later orogeny, thus making the wide part of the Piedmont in the Carolinas, but leaving no evidence in Maryland. All we know is that there was uplifting in and around Maryland, because we see the sediments that eroded off the new mountains made new rock layers to the west.

The cross section for the Acadian orogeny situation is shown in Figure 5-4A. In Maryland, the uplifted region probably was in the area of today's Piedmont. The clastic wedge that eroded off the mountains is called the Catskill Delta. This wedge is thousands of feet thick, and spreads from New York to Virginia, and west to Ohio and Tennessee. The overall size of the wedge tells us that this was a major orogeny in the Appalachians. That the wedge is thicker in New York and Pennsylvania than it is in Maryland tells us that the largest uplift occurred north of Maryland. There probably were clastic deposits east of the highlands, too, but no visible evidence of them remains in Maryland. There may have been some motion along faults, metamorphism, folding, and other effects in Maryland rocks uplifted by the Acadian orogeny, but these changes are hard to separate from similar effects which occurred in earlier and later orogenies. The orogeny apparently affected different areas of North America at different times, which makes it hard to date in Maryland. It mainly occurred in the Devonian period, but the start and end times shown in Appendix E-2 may change as new evidence is gathered.

The marine basin west of the Acadian highlands remained water-filled for a time, because the orogeny did not develop quickly in Maryland, perhaps due to Maryland's distance from the major orogeny areas. Thus, the situation shown in Figure 5-3 is still the picture for middle Devonian time. Formation of the new mountains once again bent the basin down first, resulting in the dark shales and siltstones of the Needmore through Harrell formations. Within most of the next three younger formations, the Brallier

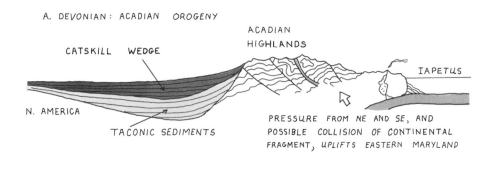

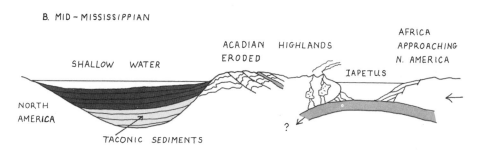

Fig. 5-4. Generalized cross sections of possible steps in the geologic history of Maryland for the Acadian Orogeny and erosion that followed.

through Foreknobs, there are variations in the sediment which tell us the water was deeper in the western part of the state, but shallower in the eastern part. Also, as we go forward in time through these layers, the shallow water indicators move farther west, testifying to the basin filling in from the east. Eventually, the deposits turn red and become nonmarine, with that change again moving west over time. In this way, the sediments had changed to the nonmarine Hampshire Formation by the end of the Devonian. This shows the Acadian orogeny had reached its peak, and sediments flowing west had filled in the basin completely to make an alluvial plain.

This plain was near sea level, with meandering stream channels and swampy areas. The Rockwell Formation is mostly alluvial plain sediments, but with some marine shale beds and shaley coals. River channel deposits of sand and gravel made most of the Purslane, along with siltstone, shale, and more shaley coal. Shallow seas then covered the area more frequently, causing the Greenbrier sediments to have more limestones and limey deposits. As with the limestone rocks of the late Silurian and early Devonian, the lack of clastic material tells us the Acadian highlands which would have supplied clastics had mostly eroded away. We have now gone through a cycle with the Acadian orogeny similar to what we did with the Taconic. You can see this in the likenesses between Figures 5-2 and 5-4. Maryland now entered a new phase in its history.

Alleghany Orogeny to the Present

The next younger rock layer above the Greenbrier is the Mauch Chunk Formation, whose nonmarine sediments show that the western part of the state was once more mostly alluvial plain. Evidently the eastern part of North America had begun to rise again to provide a source for these sediments—an orogeny that is called the Alleghany* or Alleghanian (also sometimes called the Appalachian). At this point in the late Mississippian, the orogeny only caused gentle overall tilting to the west, without major mountain building yet. It raised enough highlands, however, to provide a slow but steady supply of sediments flowing to the west through the Pennsylvanian period. Thus, this was the start of the clastic wedge of the Alleghanian orogeny, though it was formed in large measure while the orogeny developed, rather than after it had concluded. Ultimately, this orogeny was caused by the approach and collision of Africa with North America, as shown in Figure 5-4B, though the collision was only in its starting stages in late Mississippian time.

We saw alternations between shallow marine and alluvial plain environments in the Mississippian, and these changes became even more frequent in the Pennsylvanian. The developing Alleghanian orogeny kept northeastern North America, including eastern Maryland, an area of high land during the Pennsylvanian period; but areas to the south and west of these highlands experienced relatively rapid variations in sea level and shoreline position.

The area affected by the shoreline changes ran from Illinois to western Pennsylvania and Maryland. Across this region, the sediments in a column of rock often change quickly from marine to nonmarine and back again, with

---

* *Alleghany* is the correct spelling for the orogeny, *Allegheny* is the correct spelling for the plateau and the Pennsylvanian rock unit, and *Allegany* is correct for the county in Maryland. Though this may be a little confusing, it can be helpful, as the different spellings indicate what someone is writing about.

western areas showing the most marine layers. The marine layers were deposited when the area was covered by a shallow inland sea. Most of the area was close to sea level, like the Dutch Lowlands are today. Therefore, small changes in sea level could flood or drain water from large areas, as occurs now on a smaller scale on tidal flats. Also, it was easy for a river to build a delta across such a low area, and so push the shoreline out. The opposite effect happened if a stormy season washed the delta away and moved the shoreline inland again. Sedimentary conditions were variable from place to place at any particular time, because there were many promontories of land and bays of water along the irregular shoreline.

Because the in-and-out motion of a shoreline results in repeating patterns of similar sediments in the rock strata, the recurring sequence is called a cyclothem (from the Greek, meaning something laid down in cycles). Importantly, one environment that repeated in the cyclothems was a thick, coastal swamp in which great masses of vegetation accumulated and were compacted into coal. The climate was tropical in Maryland at this time, so the future coal fields were lush, swampy jungles, piling up several tens of feet of plant material for each foot of coal we find today. And the energy we get today from burning coal came to the plants from the sun some 300 million years ago. Cyclothems can include a variety of sediments besides the coal materials; in Maryland they contain shale, clay, and sandstones as well.

The general situation is shown in Figure 5-5A. The cyclothems appear in all of the Pennsylvanian formations in Maryland, but with some differences among formations. The Pottsville continued like the Mauch Chunk in being nonmarine, and has only thin coal seams. Recall that the sandstone in the Pottsville is one of the resistant sandstones that made ridges, as discussed in Chapter 3. The Allegheny was still nonmarine, but since it has more coal units, it was probably swampy more often. The sea covered more areas by the time the Conemaugh was deposited, as it contains more fossil-bearing marine shales. Finally, the Monongahela shows the best development of the coal swamps, resulting in the thickest coal beds in the state. The Monongahela contained the last of the coal beds, however, and with their end, the sea was gone forever from western Maryland.

The change from deposition to erosion over this area that had been at least partly covered by water in the past signalled the final and most active part of the Alleghany orogeny. We know this major part of the orogeny, which began in late Pennsylvanian time and continued into the Permian period, happened then because the late Pennsylvanian strata were folded by it, and they had to have been formed before they could have been folded. The

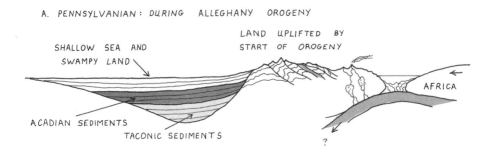

A. *PENNSYLVANIAN: DURING ALLEGHANY OROGENY*

LAND UPLIFTED BY
SHALLOW SEA AND   START OF OROGENY
SWAMPY LAND

AFRICA

ACADIAN SEDIMENTS
TACONIC SEDIMENTS
?

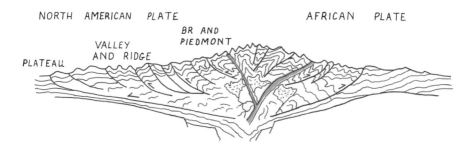

B. *PERMIAN: END OF ALLEGHANY OROGENY*

NORTH AMERICAN PLATE   AFRICAN PLATE

BR AND
VALLEY   PIEDMONT
PLATEAU   AND RIDGE

Fig. 5-5. Generalized cross sections of possible steps in the geologic history of Mary-land for the Alleghany Orogeny.

orogeny occurred because Africa had been moving toward North America from the east or southeast, and the two finally collided along what is now the southeastern edge of North America. The result is shown in Figure 5-5B. Soon after this collision, all of the major landmasses of the earth had joined together into one super-continent called Pangaea.

This collision probably generated the largest amount of pressure of all of the orogenies. It caused folding and thrust fault motion over all the Appalachian provinces in Maryland, not just the Blue Ridge and Piedmont, because the nature of the motion was different than before. The Blue Ridge and Piedmont rocks detached at their bases and were pushed as a huge thrust sheet into the sedimentary rocks of the clastic wedges to the west. This sheet in effect moved up a ramp made of the edge of the continent and travelled at least tens of miles inland, and perhaps farther. The detachment faults are shown as the long, nearly horizontal faults near the base of Figures 3-2 and 5-5A. This thrust sheet appears to extend from Pennsylvania to Alabama, and is one of the largest intact thrust sheets made of igneous and metamorphic rocks in the world. Its eastern edge is now buried under Coastal Plain sediments.

The motion of the sheet toward the northwest folded the sedimentary rocks in front of it, and created faults in them. Some ductile rock types and combinations of strata folded, while brittle layers were faulted. There were many faults, though not many are apparent on the surface in Maryland because folded layers have been stacked on top of the faults. Thus, not only did the Piedmont and Blue Ridge push over underlying rocks, but so did the Valley and Ridge and Plateau—the rocks of the whole state moved west. The combination of folding and faulting probably shortened the distance covered by the sedimentary rocks to between half and two-thirds of what it was before the orogeny. The results of this orogeny are a striking example of the power of plate motions to move and deform rocks.

We can also compare this orogeny to what we see today on the earth. Both the Himalayas in Asia and the Alps in Europe are mountain ranges caused by continental collisions. The mountains that formed in the Alleghany orogeny were not necessarily as high as those current ranges, but there are many similar structures in all three, suggesting the Alleghanian mountains were likely quite high. Also, though we think of the Appalachians in Maryland as not being far from the coast, the Alleghanian mountains would have been a mountain chain within the super-continent and not near a coast. This is the position of the Himalayas today, which lie within the landmass pieced together when India and Asia collided.

Various theories exist as to how the collision between North America and Africa actually proceeded. Particularly because the orogeny started with apparently less activity than occurred later, it has been suggested that only a promontory of land on Africa collided with North America at the beginning of the orogeny. The final orogeny then happened when full contact between the continents occurred. A promontory such as this, coming in, for example, near Maryland, might have pushed some parts of "our" Piedmont out of the way, resulting in Maryland's part of the Piedmont being less wide than it is farther south, as mentioned earlier. There is some evidence of strike-slip fault motion which could confirm this possibility. Alternatively, it has also been suggested than Africa may have approached North America at an angle from the northeast, and so struck a glancing blow at the start of the orogeny, pushing some land south, and fully colliding later. Clearly, consideration of how the outlines of the two continents might have interacted, and their angles of collision, creates many possibilities. As always, more evidence is needed to choose the correct alternative.

The Alleghany orogeny added one more metamorphic event to the eastern Piedmont rocks, giving them their final character as we see it today. With three major orogenies, it isn't hard to understand why the Piedmont rocks are in their present complex condition. This last orogeny gave the Appalachian provinces in Maryland their current overall structure, and placed them as shown in Figure 5-5B. However, the landforms were not yet as we know them today, for there was still lots of time for changes to occur before the present. In fact, the rocks at the surface today were probably thousands of feet below the surface of the Alleghanian mountains of Permian time.

Since all of Maryland was well above sea level in the Permian, most of the land was being eroded; thus, there are few rocks in Maryland from that time period. Throughout the Appalachians, however, a few closed basins were created in the orogeny, and these accumulated some sediments. These sediments make up the Dunkard Group; they are shale, siltstone, and sandstone, and are found only in a few places in the state.

Erosion continued throughout the Triassic period. Then, in late Triassic time, tensional forces began to break apart the super-continent of Pangaea. The crust broke along several parallel normal dip-slip faults that were roughly aligned with today's Atlantic Ocean shoreline, forming long, linear rift basins. This is the same type of activity that happened in the late Precambrian, shown in Figure 5-1A. Some of the faults active in the Triassic may have been thrust faults that formed in previous orogenies and were reactivated as dip-slip faults when the pressures on them were released. The rifting and resulting sedimentation probably extended into early Jurassic time, but we will refer to them as being in the Triassic for simplicity.

Most of the Triassic basins—in Maryland they are in the Frederick Valley area—were several miles wide, several tens of miles long, and deep enough to collect sediments eroding from areas around them. As with many of these basins, in Maryland they acted somewhat like a hinged trapdoor, and dropped down primarily on their western sides. Thus, the basin floor dipped to the west, and the sediments that filled the basin also dipped to the west. This is how the Newark Group formed; they were lake and floodplain deposits of rivers flowing into the basin. As is typical of nonmarine sediments, the rocks are often red in color. Notice that these deposits lie on a major unconformity, as there are no rocks with an age conforming to the period between the Ordovician and Triassic in the Piedmont. This is because uplift and/or erosion have been the main processes at work in the Piedmont since Ordovician time; any sediments that were deposited in the Piedmont in this time span were later removed. The Newark Group extends into other states in the Appalachians, and overall contains many types of fossils. A few dinosaur tracks have been found in these Triassic deposits in Maryland, and many more tracks have been found in similar sediments elsewhere in the Appalachians.

As in the Precambrian rifts, the thin and broken crust of Triassic time allowed melted basalt to rise to the surface. There probably was some general uplift of the land at this time as magma from below pushed upward. There are no surface basalt flows dating from this time in Maryland; if there ever were any, they have eroded away. But there are Triassic sills and dikes where basaltic material intruded into the Newark Group and other Piedmont rocks. This material is called diabase because it cools with a distinctive texture; it remains as low ridges today because it is a fairly resistant rock.

Unlike the Precambrian rifting which created several small continental blocks, all of the Triassic rifts became inactive before the blocks became separated, with one exception. That exception is now the Atlantic Ocean. As the Atlantic opened, we had the situation shown in Figure 5-6. The basalt that flowed into this active rift became a steady flow that today is the mid-ocean ridge, a spreading center that is injecting more material into its rift and pushing the continents steadily apart.

As the tectonic activity related to rifting ended, the eastern United States was characterized by the relative quiet of a drifting passive margin. This was the end of all of the Appalachian orogenies, and since then only erosion of rocks, with no new deposition, has gone on in the Appalachian provinces (with minor exceptions mentioned below). However, the Cretaceous and onward is the time during which the Coastal Plain province formed on the new edge of North America, which was created as the Atlantic opened. As it drifted away from the spreading center, the margin of the continent subsided. This subsided area is mainly metamorphic rocks, that is, the

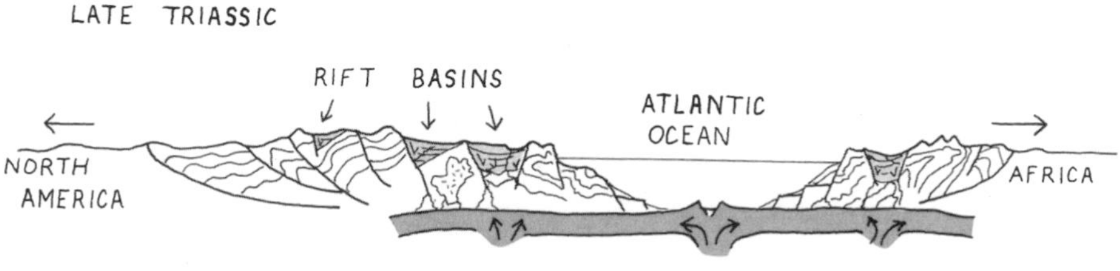

Fig. 5-6. Generalized cross section of North America and Africa in the late Triassic.

eastward continuation of the Piedmont. There may even be some terranes in these rocks that could reveal more about Maryland's history, but they are too deeply buried for us to learn much about them with our current investigative tools. The subsided rocks also contain some Triassic rift basins and their sediments, which had formed well to the east of the Frederick Valley rift.

All of these rocks are now buried beneath Coastal Plain sediments. The sediments eroding from the Appalachians have filled in and built out the shore, forming the Coastal Plain and Continental Shelf. The accumulation of sediments washing down from the mountains is the same process that occurred after the other orogenies in the Appalachians, except this time we get to see the sediments *before* they are turned to rock. In a sense, the Coastal Plain is a second clastic wedge of the Alleghanian orogeny, this time deposited after instead of during the orogeny, and east of the mountains instead of to the west. The cross section which shows these layers is the current cross section for the state, Figure 3-2.

Sea level has fluctuated while the Coastal Plain has built up, and sediments change with the environments. High sea levels resulted in marine deposits such as we now find on the shallow continental shelf, in lagoons, or along the coast; lower sea levels fostered bay or river delta deposits. When the sea was at its lowest levels, some river floodplain deposits formed, but most of the land was exposed to erosion, so deposition only occurred in places now far out to sea. These changes in conditions result in unconformities and erosion surfaces between formations in many places in the Coastal Plain.

Not only did the sediments and conditions change over time, but simultaneously there were different environments in different places. Some places were shallow water, and some were deeper; some places had fast-moving water, while other places were still. We suspect that there were some tectonic movements in the Coastal Plain area after Cretaceous time, so some areas got pushed up while others subsided. All of these effects resulted in varying sediments all over the Coastal Plain area. Thus, sediment types found today are not always continuous, which hampers efforts to trace out formations on the ground. This makes it hard for geologists to decide upon a single set of formations for the whole Coastal Plain, and accounts for different names on different local maps. The names used here and in Appendix E-3 are a compilation of those found on geologic maps of the Coastal Plain in Maryland published within the last twenty years. Other names used elsewhere or on future maps can be correlated with them. The focus in the next section is on the major environments over time, although there was variation as well.

The earliest Coastal Plain sediments, deposited by rivers, were the nonmarine floodplain deposits of the Potomac Group. Sea level rose during the late Cretaceous, and a variety of shoreline sediment types were deposited over the Potomac Group to make the Magothy Formation, which contains abundant fossil leaves. As the water continued to rise near the end of the Cretaceous period, continental shelf sediments of the Matawan Group and Monmouth Formation were added on top of the strata below. These last two formations contain many marine invertebrate fossils. Fossilized dinosaur bones are also found in the Cretaceous sediments. Entering Tertiary time, the Brightseat and Hornerstown formations, and the Pamunkey Group, all continued the marine shelf deposits, though with varying water levels, and often with unconformities between units. Then the water receded, so there are no Oligocene sediments in Maryland.

This brings us to the mid-Cenozoic, and we should look briefly back to the Appalachians at this point. As mentioned before, after the Alleghany orogeny the Appalachians could have been quite high, perhaps similar to the Alps of today. But the long expanse of time from the Jurassic to the mid-Cenozoic is enough to have worn those mountains away completely. For example, one estimate of how much erosion has occurred in the Blue Ridge is around 8,000 vertical feet. Why are any mountains left at all? Geologists are not sure, but it would seem that there has been periodic or continued slow uplift of the Appalachian region over this time. The mountains have never been high since Alleghany time, yet there has been enough relief for erosion to continue, probably at different rates at different times. The cause of this uplift is unknown, but isostatic adjustment may be involved. Also, the North American plate is being pushed from the mid Atlantic spreading center, and this pressure may move the continent vertically as well. Whatever the case may be, any uplift has aided erosion to cause the landforms to reflect the relative resistance of the underlying rocks throughout nearly all of the state.

There are many unsolved problems regarding how the Appalachian landforms have evolved; for instance, how did rivers like the Susquehanna and Potomac come to flow across the ridges of resistant rock? Perhaps cracking of the rocks during the folding processes, or the changing pattern of rock resistances that the rivers encountered as they eroded down through the rock strata helped create these pathways. Whether the accordant summits of the Valley and Ridge formed simply because the strata throughout the area are fairly consistent in thickness and relative resistance, or for some other reason, continues to be debated. Trying to deduce how landforms change over millions of years is not easy given the slow rates of change we can see. Geomorphologists are now using computer modeling, among other tools, to improve our ideas on how the processes at work on the Appalachians could have resulted in the landforms we see.

Returning our attention to the Coastal Plain, we find it was again covered by water in Miocene time. This is shown by the marine sediments of sand and clay of the Chesapeake Group, which accumulated as a landward expansion of the Continental Shelf. This group contains abundant, well-preserved fossils, including the Maryland state fossil, *Ecphora*. The quality and variety of fossils in the Chesapeake Group have made these sediments geologically world-famous. There are good exposures of these layers at Calvert Cliffs and other places along the western shore of the Chesapeake Bay.

Fluctuations in water level followed in Pliocene time, so the sediments of this epoch indicate variation from shoreline to estuary to river environments. Much of southern Maryland was covered with a variable layer (or layers) of gravel mixed with other river sediments, which covered the older sediments below. Since this layer was deposited, it has been cut through by streams so it is left on the high ground, and thus is called Upland Gravel.

The Pleistocene epoch was a time of several advances and retreats of glaciers in northern North America and other areas of the earth. Though the glaciers did not reach Maryland, they had a significant effect on the Coastal Plain because of sea level changes. When sea levels were high and covered parts of the Coastal Plain, sediments of the Columbia Group were deposited. When sea level dropped, erosion occurred, creating unconformities between the formations, and leaving the sediment layers uneven in thickness and distribution. These deposits of the Columbia Group are not very different from material being laid down today: deposits of mud, sand, and gravel along beaches, lagoons, barrier islands, and in channels. Many of the materials in the Quaternary units are from earlier sediments which have washed down streams into the estuaries and been redeposited. Thus, these deposits make up sediments near the water level of the Chesapeake Bay or Atlantic Ocean, and so are sometimes called the Lowland Deposits.

Deposition of material was not the only process in the Pleistocene that was creating something which remains as part of Maryland. Erosion made both the Chesapeake Bay and several deep canyons now under the Atlantic at the margin of the continental shelf. This erosion occurred when the sea level was low, so the combined flows of the Susquehanna, Potomac, and other rivers that currently enter the Chesapeake Bay all cut a deep valley into the Coastal Plain sediments. The ocean shoreline might have been 40 miles farther out than it is today, and the rivers extended out across the exposed Continental Shelf. Where the rivers reached the edge of the shelf, they cut canyons into it. When the water rose again, these canyons were submerged, but still exist today, as mentioned in Chapter 1. As the water rose higher, the old river valley was flooded, and thus the Chesapeake Bay was created. The Bay's dendritic pattern is the pattern we would expect, since it formed as a river valley in evenly resistant sediments.

Weathering during glacial times created another landform in Maryland. As mentioned before, the glaciers themselves were not in the state; but the glaciers occurred because the climate of the entire earth (and, therefore, Maryland) was cooler and wetter than it is today. This made weathering processes such as frost wedging more active than they are currently, especially on the tops of mountains. This created *block fields,* which are areas on mountaintops or their sides which are covered by loose angular chunks of rock, with little vegetation. In Maryland, they are made of the resistant rocks which cause the mountain to be there in the first place, and can be found in various locations in the Appalachian provinces. One is at Washington Monument State Park, as was mentioned in Chapter 2.

The most recent, and probably most active, deposits in Maryland are the alluvial deposits in the bottom of nearly every stream valley in the state. These deposits overlie the older rocks or sediments that make up the land around the streams. In some cases, the sediments in the valleys are rather thick; such deposits may have accumulated during the wetter climates of glacial times, when larger amounts of sediments would have flowed into stream valleys. When the climate became drier, reduced sediment loads in streams would have allowed the stream to cut into the deposits. The old top surface of the sediments would then be left above the river as a terrace. We do find such terraces in both the Coastal Plain and in the Appalachian provinces, and they show that erosion and deposition maintain a dynamic relation, varying over time. The Talbot Formation is a terrace deposit of the Coastal Plain, and other terraces have been found, especially along the Potomac River.

Summing up this history with a very quick look at its events, we can see the cycle that was mentioned in the plate tectonics section of this chapter: continental rifting occurred in the late Precambrian, spreading the continents apart in the Cambrian; various collisions and orogenies happened as the continents came back together until the Triassic, then the rifting occurred again, leading to the current spreading between the Americas and Europe/Africa. Will the next step, another collision to build mountains again in eastern North America, occur? Probably, though it is millions of years away. This is the classic cycle in plate tectonics which has built North America and all of the continents, and will continue to operate as long as there is heat in the earth to move the plates.

This completes our look at Maryland's long geologic history. It is amazing to realize the changes that have occurred in the land and rocks which we previously may have thought were stable and unchanging. It's also interesting to realize that all of this story was deduced merely by geologists looking carefully at the rocks, as they are our only source of information. Before we become too proud of ourselves for figuring this out, however, we should remember that the story will continually need to be updated and corrected as new facts come to light. Also, it's hard to get too big an image of ourselves when we realize the immense amount of time needed to, say, erode several successive mountain ranges, as has happened to the ground under our feet in Maryland. It may give us a new perspective on the resources of our ancient and intricately made land, which is what the next chapter is about.

Chapter Six

Maryland's Geologic Resources and Hazards

The part of the earth we call Maryland provides many resources and a few limitations for people. Though no area of the state is especially poor or rich in resources, the types found in one physiographic province may not be found in another. So, our coverage will include an overall look at each resource, followed by a look at how the provinces differ in that type. We'll look at the major resources, starting with the soil, then the water on the surface and underground, and finally, as we dig deeper, the mineral resources. There are also some hazards or limitations we should be aware of, which will be covered in the last section of the chapter.

Maryland Soils

Soil is the relatively thin layer at the surface of the earth that is composed of weathered rock and organic material from plants and animals. It is a vital resource for people, because our lives depend on growing food from the soil. The soil in one location is different from that in another for several reasons, but some overall characteristics will be similar, and we'll look at these features first.

Most soils develop several layers, called horizons by soil scientists, as shown in Figure 6-1. The boundary between any two of these layers is rarely sharp; instead, one horizon grades into the next one. The top layer, the O horizon, is the organic material called humus—leaves, old plants, and sticks—that covers the soil, and is beginning to become part of it. The next layer, the A horizon, is the topsoil. Here, roots, soil-dwelling animals of all sizes, and decaying humus are mixed with the mineral soil towards the top of the horizon. Lower in the horizon is a leached area, where certain minerals and clays have been washed out by rainwater as it moves down through the soil. Below that, the B horizon is where these leached materials accumulate, making the layer called subsoil. The C horizon consists of the starting material of the soil in advanced stages of weathering; this starting material

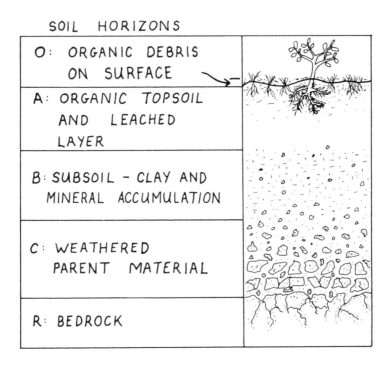

Fig. 6-1. Typical characteristics of a soil profile in Maryland.

may be bedrock or other material, and is generally called *parent material* as in Figure 6-1. If the parent material is the bedrock, it probably would be soft and breakable in your hand if it is in the C horizon. In the Maryland Piedmont soils especially, there may be a thick layer of this soft, thoroughly weathered rock material, which geomorphologists call *saprolite*. In the Piedmont, the saprolite may be 50 to 60 feet thick on average, and is several hundred feet thick in some places. Below the C horizon is relatively unweathered parent material called the R horizon, on which the soil rests.

The thickness of each of these layers varies from place to place in Maryland, even over short distances. Soil tends to be thin on the tops of hills and on steep slopes. Resistant rocks also tend to have thin soil covers, because the rocks weather into soil so slowly. Soil is usually thicker on gentle slopes and in valleys, where it can form and not be washed away.

The nature of the soil in a particular location depends on the rock type from which it has developed, the climate, and a number of other factors. Sandstones can produce a sandy soil, while shales may produce a soil with more clay. Granite, schist, and gneiss can make a mixture of sand and clay called *loam*, and probably would also contribute noticeable amounts of mica. Soil on basalt or gabbro would probably be clay, colored red by the oxidized iron minerals of the rock; indeed, even small amounts of rust-colored iron oxides are a major cause for the brown color of most soils. Limestones are likely to produce a lime-rich clay soil.

However, soils on two different bedrocks may be similar if they develop for a long time under similar conditions, because many of the same chemical reactions are going on in both. Meanwhile, variations in slope, moisture, plants, and mixing of soil types caused by erosion can make for variable soils even on the same bedrock. When scientists classify the soils in different places, performing what is called a soil survey, they find many different types across the state. Surveys also may find significant variations in soil types even in a relatively small area such as an acre.

Soil surveys group soil types into classes which indicate appropriate uses for the land. Some areas are particularly good for farm cultivation. Other soils and lands are suitable only for limited cultivation, but may be used successfully for less intensive purposes, such as grazing animals on the land. Some soils are best left with a forest or other natural cover because they are on land that is too steep, marshy, rocky, or sandy for other uses. It is important that we pay attention to such limitations so that we can use the soil resource properly, and still preserve it for future generations to use.

A good illustration of how people can affect soil and erosion is found in the sediments which have accumulated in the bottom of the Chesapeake Bay. Measurements of these sediments indicate that the rate of sedimentation (that is, how fast sediment has washed in) increased once people settled in Maryland and the rest of the Bay's watershed and began to make changes—clearing forests, farming, and quarrying stone. After soil conservation practices became more common in the 1930s, the sedimentation rate decreased somewhat. This measure of people's effects on the land and the Bay is so sensitive that we can even find in the sediments a layer of charcoal and ash that was washed in after the Baltimore Fire of 1904. Clearly, we are capable of damaging our soils if we are not careful with them.

Soils of the Provinces

Coastal Plain soils tend to directly reflect the sediment on which they have developed, since these sediments are much like soil anyway in many places. Therefore, some contain more clay, and others are sandy. Most are a mixture of materials excellent for cultivation, and so support the large farms of the province. Occasionally, these soils contain clay layers that do not allow water to pass through them easily, and thus can be said to be relatively impermeable; they are less useful for farming than other Coastal Plain soils. Because of the low topographic relief of the Coastal Plain, some areas are too wet for farming. Soil surveys are useful in such cases to determine optimum uses—including leaving an area in its natural state.

In the Piedmont, soils on the schists and gneisses contain both clay and sand, providing a texture suitable for farming. They are naturally fertile because they contain a variety of minerals. However, since the rocks are moderately resistant, the soil is thin in some locations, and needs to be built up with organic material such as manure and mulch. Exceptional areas in the Piedmont are the serpentine barrens, such as Soldiers Delight (see Figure 3-10). The bedrock there, serpentinite, does not have the variety of minerals that other rocks do, so this soil is less fertile chemically than others in the Piedmont. As a result, these barrens have developed forests and grasslands with some plants that are uncommon elsewhere (e.g., certain types of oaks), and many plants that do not grow as large there as they do in other soils.

In the Piedmont and other Appalachian provinces, the soil types on similar rocks are similar to each other, too, no matter where they are found. The soils on limestones tend to be moderately permeable, with good sub-surface drainage due to the dissolving of the rock. The red rocks that occur in parts of the Appalachian provinces result in rather red soils, more sandy and permeable in the red sandstones than in the red shales. The resistant sandstones break into pieces that remain in the thin soils around them, producing rocky or cobbly soils. The shales of the valleys weather to moderately permeable soils, but often need fertilizing because they are not very fertile by themselves.

Overall, the soil types in Maryland have given the state a good agricultural economic base. Different soils have different needs and limitations in the varying parts of the state, but with care our soils will continue to serve us well.

Water Supplies of Maryland

Some have said that our planet is better named Planet Water than Planet Earth because there is so much water on and near the surface. Indeed there is a lot of water, and Maryland has its share. Water as a resource is different from most in that it moves around constantly, and is naturally recycled in time spans which are shorter than geologic length. This overall process, called the hydrologic cycle, is shown in Figure 6-2, and described below.

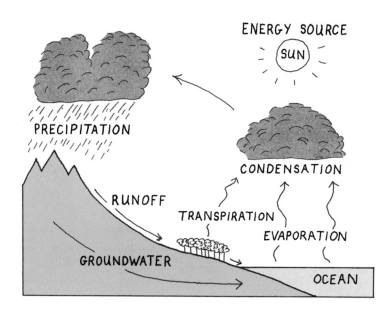

Fig. 6-2. The hydrologic cycle.

The heat of the sun constantly evaporates water from the oceans. In the atmosphere, it condenses to make clouds, which then produce precipitation as rain or snow. This water either runs over the surface of the land or runs underground, eventually reaching the sea to maintain the cycle. Some water also takes a shorter route by evaporating from the land or from plants (in a process called transpiration), without ever reaching the sea.

The runoff water on the surface flows, of course, into rivers and streams. Less obviously, a large amount of water is also stored in and moves through the ground itself. Contrary to some popular ideas, rainwater that has percolated into the ground does not usually flow in underground streams. Most of the time, it moves through small gaps and cracks in the rocks and so permeates the whole rock, somewhat like water flowing through a sponge. The area of the ground which is filled with water is called the *zone of saturation*. Above this area are soil and rock with air in the open spaces, so this part is called the *zone of aeration*. The surface between these two zones is the *water table*, as shown in Figure 6-3.

The water table rises and falls with the surface of the land, though in a more gentle and subdued way, which is also shown in Figure 6-3. Whenever the water table intersects the surface, water flows out of the ground. This

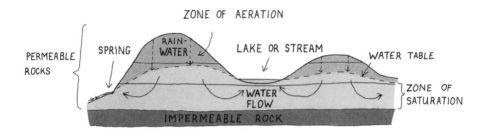

Fig. 6-3. Cross section of land showing the water table, and its relation to surface features.

could create a spring, or feed water into a stream from the stream bed. The elevation of the water table also varies with the amount of rainfall. Thus, in a dry period the water table may drop so it no longer intersects the surface; then springs will cease to run, and streams begin to lose water into the ground. In this way, surface water may disappear though there is still some water in the ground.

The ability of water to flow through the ground depends on the materials in the earth. Sand and gravel have many small openings among the grains, and so are said to be very *porous*. These openings are well connected, so water can flow through them easily; thus we say these materials are *permeable* also. A layer of these or any other material which can hold and move water is called an *aquifer*. On the other hand, clay, though fairly porous, is rather impermeable, and so not much water is carried in clay layers. In fact, clay layers can serve as barriers to water flow, holding water above or below them, or trapping it between successive layers.

Sedimentary rocks generally follow the same pattern as the loose sediments that they're derived from with regard to groundwater flow. Sandstone is usually permeable (unless its pores are filled with cementing material), but shale usually isn't. Limestone has interlocking grains, and so has few pores for water to move through. However, water flowing through cracks in limestone will dissolve the rock and so widen the cracks; thus, fractured limestone can carry large quantities of underground water. Igneous and metamorphic rocks are neither porous nor permeable, and so only carry water if they have cracks. But it turns out that the forces that create igneous and metamorphic rocks often create regular patterns of cracks called *joints*, and these may carry enough water to allow, for example, for residential wells. In marble, these joints may be widened by water flow, as occurs in limestone; thus, marbles may carry more water than some other metamorphic rocks. Examples of some of these rocks in Maryland, and the water flow through them, is shown in Figure 6-4.

Before looking at how water resources vary in each province, we can look at the statewide situation. Maryland receives about 9 trillion (that's 9,000,000,000,000) gallons of water in precipitation each year, plus 11 trillion gallons flowing into the state in streams from adjacent states. Of that, about 6 trillion gallons are lost to evaporation and transpiration from plants. That leaves about 14 trillion gallons available to use. Some uses—

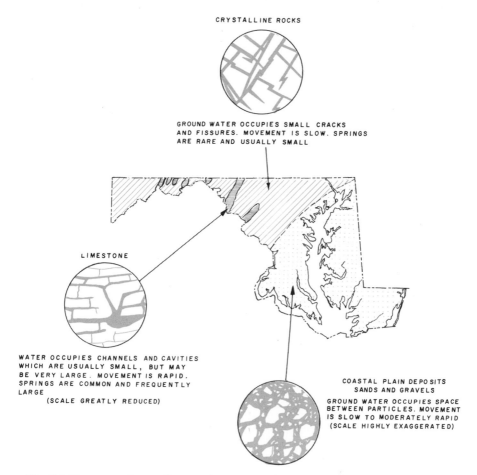

CRYSTALLINE ROCKS

GROUND WATER OCCUPIES SMALL CRACKS AND FISSURES. MOVEMENT IS SLOW. SPRINGS ARE RARE AND USUALLY SMALL

LIMESTONE

WATER OCCUPIES CHANNELS AND CAVITIES WHICH ARE USUALLY SMALL, BUT MAY BE VERY LARGE. MOVEMENT IS RAPID. SPRINGS ARE COMMON AND FREQUENTLY LARGE
(SCALE GREATLY REDUCED)

COASTAL PLAIN DEPOSITS SANDS AND GRAVELS

GROUND WATER OCCUPIES SPACE BETWEEN PARTICLES. MOVEMENT IS SLOW TO MODERATELY RAPID (SCALE HIGHLY EXAGGERATED)

Fig. 6-4. How groundwater flows and is contained in rocks in various areas of Maryland. Source: Maryland Geological Survey Educational Series No. 2, Water in Maryland: A Review of the Free State's Liquid Assets.

such as nonpolluting cooling water or hydroelectric power—simply return it to the cycle without much change; these are called nonconsumptive uses. Other uses are called consumptive because the water quality or quantity is lessened; this includes most home uses, and evaporative uses like air conditioning. Consumptive use was about 1.2 trillion gallons in 1980 and is predicted to be about 2.5 trillion gallons by 2010. Importantly, this use is increasing on a per-person basis—from 95,000 to 120,000 gallons per person per year from 1980 to 2010 (of course, much of this is industrial, not individual)—so we need to be aware that the supply of good water is not an endless one. Though we seem to have an excess of water, having it at the right place at the right time, and of the right quality, takes careful management and everyone's efforts to prevent waste or contamination of water.

Water in the Provinces

The main source of water in the Coastal Plain is groundwater. Because the province is underlain by unconsolidated sediments, precipitation usually sinks in easily. There, the water moves through and is stored primarily in the sandy layers that are the aquifers for the region. Wells into these layers provide more than 120 million gallons a day to homes, farms, and industry in this area. The water in the aquifers is generally of good quality, though in some places the groundwater picks up an undesirable amount of iron, calcium, or magnesium as it moves through the sediments.

For all the provinces west of the Fall Zone, that is, the Appalachian ones, the main water source is surface water, not groundwater. The water for the larger cities comes from various man-made reservoirs, which pond the water for the times when rainfall and therefore surface runoff is low. All of the numerous lakes shown in Figure 3-10 are examples of these reservoirs, as the concentrated population in this area demands lots of water. The relatively deep, dendritic valleys of the Piedmont, which often have formed where there are cracks in the underlying rock, are the "holding tanks" for these reservoirs.

Wells in the Appalachian provinces usually only produce enough water to supply domestic needs, not enough for cities or industry. In the metamorphic rock areas of the Piedmont, groundwater is found in the saprolite layer and in joints in the rock below. Because the saprolite stores water somewhat like a sponge, wells drilled in areas of thick saprolite are more reliable and show only small fluctuations in their water levels in dry times. Marbles in the Piedmont often have more water than wells in other rocks because joints in the marbles are larger, widened by the water itself. In general, wells in flat or valley areas tend to yield more water than wells on hills or slopes, though in the end the rock type and other geological factors are more important than the landforms in determining the productivity of a well.

In the Blue Ridge, most water comes from springs. Limestone in the Hagerstown Valley yields good water supplies from wells if the well is drilled or dug into a fractured area in the rock. Wells in the impermeable, well-cemented sandstones in the western part of the state can be nearly useless for water, but fractured shales and less-cemented sandstones do supply some wells with adequate water. A groundwater availability map would look much like Figure 6-4: high availability in the Coastal Plain, moderate in the limestone and marble areas, and low in the crystalline rocks.

Looking at the state as a whole, Maryland's water supply is good, both in amount and in quality. It does require protection, however, as it is vital to our survival.

Mineral Resources

Mineral resources depend directly on the underlying bedrock or sediments in an area. Because Maryland has a quite varied geology, it has a rather wide variety of mineral resources. Some of these were important to the early economy of the state, but are less so today because better sources have been found elsewhere. Other resources are used extensively today, and active quarrying and mining for these materials goes on throughout the state. The resources available from each area of the state are different, depending on the geology, so we will examine them by province, grouping similar provinces together.

Coastal Plain Province

The Coastal Plain serves as a resource area for the materials of which it is made: sand, gravel, and clay. Sand and gravel are important as construction materials and accounted for about 27 percent of the value of all nonfuel mineral production in Maryland in 1987. This material came especially from the Coastal Plain deposits, since only sorting of the material is needed, without extensive crushing required. The Patuxent Formation and Brandywine deposits are the main sources of Coastal Plain sand and gravel, and the Upland Gravels of southern Maryland.

The other important resource of the Coastal Plain deposits is clay. In the Arundel and Patapsco formations, and the Marlboro Clay, are clay deposits that are used for making bricks and terra cotta items. Clay for pottery, called ball clay, is found in these formations also.

Other materials of the Coastal Plain, important in the past, are no longer used. Iron was produced from ores in the Arundel Formation from the early 1700s to the early 1900s. Thousands of tons were produced annually, for iron was important both to early settlement and development of the state, and for military uses in the Revolutionary War and the War of 1812. Kaolin, or white clay, used in ceramics and as filler or coatings for various materials, was produced in large quantities in the past in northeastern Maryland. Fire clay, from which fire bricks that can withstand the heat inside kilns and furnaces are made, once came from Coastal Plain deposits. The Calvert Formation was once the only source in the United States for diatomaceous earth, until other deposits were found in California. This material consists of the hard siliceous shells of microscopic plants called diatoms, and is used as a filtering material or as an abrasive.

Piedmont and Blue Ridge Provinces

We'll cover the Piedmont and Blue Ridge provinces together because the latter is a small province, and because their rocks and resources are similar. These areas of the state contain the hard metamorphic rocks used especially for building stones. The supporting structure in most buildings today is steel and concrete rather than stone, but stone is still used as facings for decorative effects. The wide variety of rock types in the Piedmont provides many different ornamental possibilities.

Several granites, including the Port Deposit, Ellicott City, Woodstock, and Guilford granites, have been used at various times for building purposes. Today, these rocks can still be quarried when there is sufficient demand. The Weverton Quartzite has also been used as a building stone, for example, in the monument at Washington Monument State Park and in several structures at Gambrill State Park. Several quarries for roofing slate have operated in the Piedmont, but none do today. The Setters Quartzite is frequently seen as a facing stone or flagstone, because clays in the original sandstone have been metamorphosed into mica layers, which allow the rock to be cleaved easily into flat-sided pieces.

The Cockeysville Marble was used in the Washington Monument in Baltimore, the National Capitol in Washington, and other major buildings in Baltimore and the eastern United States. It is also the rock used to make many of the white marble steps of Baltimore's row houses. Serpentinite's attractive green color has led to its use as a decorative rock known as "green marble." Finally, though not metamorphic rock, the sandstones of Triassic age in the western Piedmont (called "Seneca Red Stone," or "brownstone," after its colors) have been used in many places. They can be seen as part of the Patowmack Canal around Great Falls of the Potomac, the Chesapeake and Ohio Canal, and the Smithsonian Institution's castle in Washington, DC. They also make up some of the brownstone houses in Baltimore and Washington.

More than 30 million short tons of crushed stone represented about 44 percent of the value of all nonfuel mineral production in Maryland in 1987. The Texas Quarry, in Cockeysville Marble 15 miles north of Baltimore, was one of the ten largest producers of crushed stone in the United States in 1985. In the Piedmont, the major rocks used for crushed stone are the marbles, gabbros, serpentinites, and some limestones. Large amounts of crushed stone are used in construction, especially for roads and as fill material.

Piedmont rocks are also used in other ways. Limestones in Frederick County have been used to make various components of cement for many years. Some clays from weathered Triassic shale are used to make bricks today, and clay resulting from weathering of metamorphic and igneous rocks has been used for this purpose in the past. A shale layer in the Frederick Limestone serves as a source of lightweight aggregate, which is used as a filler in concrete to decrease its density.

The processes that make metamorphic and igneous rocks also lead to concentration of metallic minerals. A variety of metal ores found in the Piedmont were important in the early economic development of Maryland and the United States. Mining of these materials has dropped as richer and larger deposits have been found elsewhere, but it is still interesting to note what ores are here. Iron was removed from near Catoctin Mountain from the late 1700s to the early 1900s. Copper was mined at a number of places, and one mine produced nearly $1.8 million worth of ore between 1861 and 1886. Relatively small amounts of lead, gold, and silver have been mined in the Piedmont, and these metals were often found in the same place. Chromite, which is the ore of chromium, was mined from the serpentinite formations, and made Maryland the leading chrome ore producer in the world from 1828 to 1850. Small amounts of manganese, molybdenum, and titanium minerals have been found, but only the first of these was actually mined.

The Piedmont area has also produced a few other nonmetallic materials. Feldspar for ceramics was mined from pegmatite dikes during the first half of the twentieth century. Quartz, sold under the trade name "flint," came from pegmatites and other veins, and was used in ceramics, abrasives, and as filler. Finally, soapstone from the Piedmont, in the form of a schist made mostly of the mineral talc, has been quarried as slab stone, since it is easily cut. Soapstone has also been ground to use its talc as filler in paper, paint, rubber, and other materials.

Valley and Ridge, and Allegheny Plateau Provinces
Similar rock types underlie both the Plateau and the Valley and Ridge provinces, and so their resources are alike in some ways. We'll start our look

at their resources, however, with a way in which they are different: nearly all of the coal is found in the younger rock formations of the Plateau, with only a small amount farther east, as shown in Figure 6-5.

Coal mining became an important industry in Maryland in 1842 when the Baltimore and Ohio Railroad reached Cumberland, thereby providing cheap transport of the coal to large markets. Production peaked in 1907 at more than 5.5 million tons, but has been less since then. Most of Maryland's coals are semibituminous, making them intermediate in the hardness range of coal types, and of a type which produces more heat per pound than other varieties. The thickest coal layer in Maryland—14 feet—is at the base of the Monongahela Group. Much of this layer has been exhausted, so work now is focused on thinner seams which require more effort per ton of coal removed; this is one reason why coal production has decreased overall. Today, most coal is extracted by strip mining, where coal is removed at the surface, after the overlying material has been removed. Though the process is getting more expensive as the deeper coal is mined, there are large reserves to maintain the industry for a long time to come.

Natural gas also has been found mostly in the Plateau region, as shown in Figure 6-5. Each of these gas fields is on an anticline, or, in the vicinity

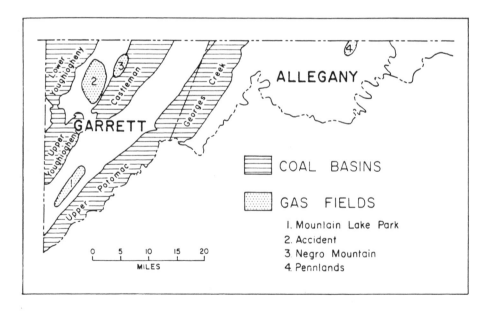

Fig. 6-5. Coal basins and gas fields in western Maryland. Source: Maryland Geological Survey Bulletin 19.

of Accident, on a dome. Sometimes, gas collects in a permeable rock in the high part of the fold, like hot air rising to the high point inside a peaked roof; the gas is then held in by impermeable overlying layers (often shale). In the dome at Accident, the gas has collected in cracks that formed during thrust faulting of the underground Oriskany Sandstone. The gas flows out where wells are drilled in this rock, some of them 5,000 feet deep. Though no new wells are being drilled at the present time, total gas production since about 1949 has been 45 to 50 billion cubic feet.

The Valley and Ridge and Allegheny provinces are also similar in their nonfuel resources. Limestones in these provinces are put to two uses: first, as a crushed stone for building purposes in both provinces and, second, as the major ingredient in cement-making. Another material found in these provinces is clay. Weathered clay from the Martinsburg Shale goes into bricks in Washington County. The underclays associated with some of the coal beds of the Plateau have been used as fire clays in the past.

Have we found all of the geologic resources available in Maryland? Probably not. As we learn how mineral and other resources are concentrated in the processes of plate tectonics, we learn of places to search for these resources where we hadn't thought to look before. Because Maryland has had many plate tectonic events and environments—rifts, island arcs, collided continents, and passive margins—there are many possible locations where resources might be discovered, once we know in what type of place to prospect. Thus, the general understanding of the geology and geologic history of the state can lead us to resources we need, because it tells us where to look. No one can guarantee how many new untapped treasure troves we might find in Maryland, if any, but increasing geologic knowledge is the best direction we can take to better our chances of making new resource discoveries.

Geologic Hazards and Limitations

"Nature, to be commanded, must be obeyed."

Francis Bacon

"Human beings live by geologic consent—revokable at any time."

Will Durant

People in Maryland are fortunate to have many geologic resources and few serious geologic hazards in the state. However, we should remember that our lives depend on our planet, as expressed in the two quotations above, and we ignore the earth's natural processes only at our peril. Therefore, while looking at the good things the earth provides, we must also look at aspects which could harm us, especially if we are careless.

When people consider geologic hazards, they are likely to think first of volcanoes and earthquakes; do we need to worry about these in Maryland? Obviously, there are no active volcanoes in Maryland, and practically no possibility of one forming anytime soon, so this needn't be a major concern. While we mostly think of earthquakes occurring in California or other distant places, certainly Maryland has many old faults that could become active if pressures on the crustal plates made them do so. There are usually a few small earthquakes in the Appalachians each year, and very infrequent but major quakes in eastern North America, so the potential for a damaging earthquake does exist. However, since Maryland is located in the middle of a crustal plate, and few such places have a much lower possibility for a significant earthquake, we are in about the safest spot we can be. Designing structures to withstand earthquakes, and including their possibility in disaster preparedness plans is not a bad idea, but the likelihood of a major occurrence here appears to be small. There are, however, many less spectacular but more likely geologic problems that need our attention.

One limitation that applies to all of Maryland is that our rich resources of surface and groundwater must be protected. Wasting drinkable water, or contamination and pollution of it, can reduce an abundant resource to one that is hard to obtain. Contaminating chemicals dumped on the ground can be carried by surface or groundwater to the reservoirs we depend on in the Appalachian provinces of the state. Similarly, pollution can be carried into the groundwater of any area of the state and so render wells useless for drinking water.

The permeability of both the sediments in the Coastal Plain and the limestones of the other provinces makes them particularly vulnerable to groundwater contamination. On the other hand, impermeable or unfractured crystalline rock can act to contain pollutants, though this might also have the unfortunate effect of concentrating the pollution. Because groundwater flow is rather slow, polluting chemicals may take some time to thoroughly contaminate the groundwater in an area, but then can remain there for long periods. Toxic chemicals generated by industry or individuals are a growing problem in our water supplies, and efforts now are beginning to be focused on eliminating their production in the first place. Such efforts might require all of us changing our habits, but may be necessary to assure we have clean water available.

Another limitation applies to the Atlantic and Chesapeake Bay shore areas, and the Coastal Plain as a whole. Though we may feel the locations of shorelines are constant, a geologic perspective shows they are very changeable. Evidence suggests that sea level is currently rising—somewhat less than one foot in the last 100 years—which means that the shoreline will move inland; therefore, mankind's structures near the beaches are in jeopardy. Since the Coast Plain is of such low relief, even small rises can cause clear changes, and these become especially important when tides or waves are high due to storms or other special conditions. The tidal shore-lines of Baltimore and Washington would be affected by these changes, too. Even without a rise in sea level, beaches in general and barrier islands in particular are changeable features. They do not last very long in any one form, and may be washed away completely by a large storm or stormy season. Ultimately, barrier islands such as those along Maryland's Atlantic shore will be flooded, washed away, or at least moved farther inland. Efforts in the past to prevent such changes have generally succeeded, often at significant expense, for only relatively short periods of time, but have ultimately failed. Thus, it is sensible to question whether extensive building is appropriate for such areas, if beaches and barrier islands are geologically temporary features.

People may even be contributing to the rising sea level, as our use of fossil fuels such as coal and oil increases the carbon dioxide amounts in the atmosphere. We also increase carbon dioxide because we cut down the forests which would help to use it up as the plants grow. This carbon dioxide increase may cause a significant warming of the earth (known as the greenhouse effect), melting glaciers worldwide, and adding more water to the oceans. Predictions vary on how much this will occur, and if it will occur at all, but even a rise of 1 to 2 feet in the next 50 to 100 years would have noticeable effects on shoreline locations.

We can't be sure at this moment whether the greenhouse warming will really happen, but it may be time to take steps to be sure it doesn't. This situation, though not strictly geologic, shows how the land we live on is affected by what we do, and knowledge of what changes we may be causing is (or should be) important in determining our future actions.

Even people living in areas away from the ocean or the Chesapeake Bay must deal with problems of excess water, in the form of floods. Maryland does not have a particularly rainy season like some areas of the earth, but we do get heavy rainfall during thunderstorms, and especially when a hurricane strikes the state. Exceptionally high runoff can occur in some of the narrow valleys of the Appalachians, where a cloudburst can be concentrated into streams all at once. As examples of the effects of this, the large variations in flow in three different rivers in the state are shown in Figure 6-6. Areas covered with buildings, roads, and parking lots also generate high stream flows during rains because little water can soak into the ground. When high runoff occurs, people will be better off if they have considered the possibility of floods beforehand, and not used land near rivers and streams for building. Agreeing to limitations on land use is one way of using our geologic knowledge to make our lives easier and safer in the long run.

Another limitation relating to the land applies to any hill. On a slope, all material is moving downhill due to gravity and water erosion. On gentle slopes, the movement is called *creep* and is very slow, but it goes on all the time. Thus, homeowners find retaining walls need replacing from time to time as movement of the slope pushes them over. When water is added to the picture, we can see more soil erosion occurring on slopes, especially on ground bared for construction or cultivation. This causes loss of soil and pollution of streams with excess sediment; the only way to reduce it is to spend more effort and money to keep the ground covered with plants or artificial protection. Excess water absorbed into a slope can lead to soil and rock sliding as a unit of ground moves downslope; this type of motion is called a *slump*. The slopes in the Plateau may be especially subject to slumping, which can damage property and disrupt roads.

The most extreme case of material moving downslope is a landslide. Landslides are a normal part of the erosion processes that have been at work on the Appalachians for more than 200 million years. The Valley and Ridge is notably susceptible to landslides, particularly during exceptionally heavy rains. We cannot stop the landslides, but we can be aware of their possibility in sloping areas, and not build where they are likely to occur. We also need to be careful not to undercut slopes, which would create a landslide-prone spot. All of these examples indicate how important it is to be aware of the natural processes at work on the land, and to plan and act accordingly.

We can further compound geologic hazards if we make changes to the earth. Certain operations are important to our way of life and economic needs, and no one proposes stopping them. At the same time, they have often caused problems of a geologic nature.

One process that can produce significant changes is mining in its various forms, including, in Maryland, rock quarrying and coal mining. Quarries can disrupt or pollute groundwater supplies if not properly managed. Mines especially can and have caused acid runoff water as

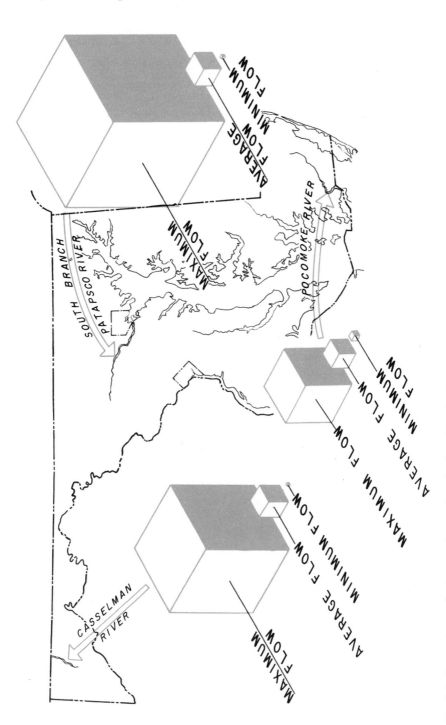

Fig. 6-6. Comparisons of volumes of water flow for three Maryland rivers. Source: Maryland Geological Survey Educational Series No. 2.

chemicals in rocks are brought to the surface at a faster rate than would occur naturally. And, of course, there is the physical disturbance of the ground, resulting in waste piles which are unsightly and sometimes dangerously unstable.

Another process that changes the land is farming. Farming can bring pollution and changes to streams, if runoff water from fields carries agricultural chemicals or soil as added sediment.

These problems can often be overcome, with effort and money, and many people now agree these costs should be borne by those benefiting from the products that result from the land use. This means higher prices for coal, crushed rock, food, and other resources. But that seems better than trying to ignore their geologic costs, and passing these costs on to future generations, who received no benefits from the materials. Thus, we need to see the geologic consequences of our actions, to analyze and compare the good and bad points of what we are doing.

Radon gas is a geological hazard that we have only become aware of recently, and on which studies continue to be done. Certain materials in the earth are radioactive; that is, they give off energy spontaneously as changes occur to the structure of their atoms. Radon is a radioactive element that is a gas when found at the earth's surface. This gas is released from rocks in the earth, some rocks containing and releasing more of it than others. Generally, the rate of release is very slow, so the gas is dispersed by the wind and does not build up in any one place. However, some radon escaping from the rocks will naturally enter buildings built over these rocks. As long as the building is well ventilated, the radon will leave the building and mix with the atmosphere as though the building weren't there. But if the building is closed, as are most basements and whole buildings during the winter, the radon may accumulate. Evidence suggests that living in a building (home or office) that has high radon levels may be hazardous to health, because the radioactive energy given off by the gas may damage our bodies as we breathe it in.

The amount of radon in a building depends partly on the geology of the area. Metamorphic and igneous rocks, especially those of granitic composition, as well as some sedimentary rocks like black shale, may be likely to release significant amounts of radon. Thus, a few buildings in the Piedmont of Maryland and other states have shown especially high levels of radon, and studies continue throughout the Appalachian states on the relationship between radon and the underlying rocks. The amount of radon in a building also depends on the building itself, particularly on how well it is sealed from outside air. So, this is another problem that illustrates how geology interacts with our actions and decisions—in this case, where and how we construct our buildings, and then how we use them. It is not really a new problem, though it has become more noticeable lately as we have built structures that have less outside ventilation, for the praiseworthy reason of saving energy. Rather, this is a problem about which our expanding science has recently informed us, that we can now work on solving and correcting. Knowledge of this geologic situation warns us to do something before a more serious problem—poor health—appears.

All of the hazards and limitations given above indicate that geologic processes and geologic understanding can play an important part in many plans and decisions that mankind makes. A plan to build a highway, put up a building, drill a well, start a farm, construct a breakwater, or rebuild a beach is not really complete until its geologic aspects are considered. In our modern world, often cut off or protected from the actual processes of the earth, we may feel we can do what we want and not worry about these mechanisms. But the facts are otherwise: our soil, water, energy needs, health, and even the stability of the very ground we walk and build on depends on our interactions with geologic and other natural workings. And some of our current problems—polluted groundwater, erosion at beachfront buildings, or excessive soil erosion, to name a few—are in part caused by our past disregard for geologic processes. These processes won't go away if we ignore them. A better approach to living on the earth is to understand as much as we can about it, and use that knowledge to determine our actions.

Appendix A

Summary of the Features of the Physiographic Provinces of Maryland

PROVINCE	TOPOGRAPHY	PREDOMINANT ROCK TYPE	ROCK STRUCTURE	PREDOMINANT AGE OF ROCKS
Coastal Plain	flat, low relief; more hills in western shore area	sediments not consolidated into rocks	dipping gently toward ocean	Cenozoic & late Mesozoic
Piedmont	rolling hills & stream valleys	metamorphic & igneous; sedimentary in Frederick Valley area	complex folds & thrust faults	Precambrian & Paleozoic
Blue Ridge	mountain ridges	metamorphic	broad anticline with faults & complex subfolds	Precambrian & early Paleozoic
Valley & Ridge	long mountain ridges of resistant rocks alternating with linear valleys cut in nonresistant rocks	sedimentary	anticlines & synclines, & faults	Paleozoic
Allegheny Plateau	high plateau with deep stream valleys and some ridges of resistant rocks	sedimentary	uplifted, warped by broad folds	Paleozoic

Rock Types Found in Maryland

Igneous Rocks

These are rocks which have cooled and solidified from magma, which is a mass of melted minerals in the earth. If the rock cools slowly underground, it usually has individual mineral grains large enough to be seen, and is labelled coarse-grained. We also say such a rock is *intrusive* because it has been injected into or intruded into the existing underground rocks. On the other hand, if magma flows out on the surface as lava, usually at a volcano, it will cool relatively quickly to make an igneous rock. Individual mineral grains won't have time to become large enough to be seen, and the rock will be fine-grained. Volcanic rock such as this is said to be *extrusive*, as it has been squeezed out or extruded onto the surface.

Igneous Rock Types in Maryland

Granite - coarse-grained, mostly made of felsic minerals.
Rhyolite* - fine-grained, mostly made of felsic minerals.
Gabbro - coarse-grained, mostly made of mafic minerals.
Basalt* - fine-grained, mostly made of mafic minerals.
Ash* - powdery deposit of fine, airborne particles ejected from volcanoes.
Tuff* - ash deposit which was hot enough to be welded together into solid rock at the time it erupted from a volcano.
Pegmatite - very large-grained, usually made of felsic minerals; nearly always occurs in a small rock body such as a dike, which forms when molten rock intrudes and solidifies in a crack in an existing rock.

*This rock type doesn't occur in Maryland in unaltered form, but it is the starting material for other rocks that do occur; see note on metamorphic rocks listed below.

Diabase - medium- to fine-grained mafic rock similar to basalt but with a distinctive microscopic texture; usually forms in intrusive bodies which solidify in cracks in rocks, called sills if near horizontal, or dikes if near vertical.

Sedimentary Rocks

These are rocks formed by the depositing and/or cementing together of sediment, which is material that has come from other rocks. If the sediment is particles (clasts) of other rocks which have broken apart due to the action of weather and running water, it is described as clastic, and produces clastic sedimentary rocks. Or the sediments can be plant or animal remains or products, or may be deposited due to the evaporation of water; these produce rocks which are classed as nonclastics. Often, sedimentary rocks are a combination of sediment types, and usually are deposited in horizontal layers called strata or beds. The sediments collect at or near the surface of the earth, but then must be buried in order to turn into rock.

Sedimentary Rock Types Found in Maryland
Clastics:
  Conglomerate - aggregate made of rounded rocks and pebbles of other rocks cemented together with finer-grained material; may look like concrete.
  Breccia - made of angular (broken or sharp-pointed) pieces of other rocks cemented together with finer-grained material.
  Sandstone - made of sand grains cemented together, usually made up mostly of quartz.
  Graywacke - made of sand-sized grains, containing a mixture of quartz particles and rock fragments, cemented together by clay.
  Siltstone - made of slightly gritty dust or mud particles.
  Shale - made of very small grains of clay, often smooth on the surface of the rock; forms in distinct layers and often breaks along these layers.
  Mudstone - made of clay, but without the obvious layering found in shale.
Nonclastics:
  Limestone - made of calcite; comes in many textures and colors.
  Dolomite - made of the mineral dolomite; similar to limestone.

Chert - very fine-grained form of quartz; a broken piece is very smooth to the touch.

Coal - made of compressed and altered plant remains, with some clay.

Sediments are often mixed together, and so are the rock names. For example, a shaley limestone would be limestone with clay mixed in. Or a limey sandstone would be a sandstone with some calcite in it (also called calcareous sandstone).

Metamorphic Rocks

These are rocks in which the minerals and textures have been changed from their original condition by high heat and pressure, but not enough to melt them. Examples of the kinds of changes that occur are rearrangement of elements to make new minerals, movement of minerals around to make layers, and enlargement of individual grains. If a rock has changed drastically from its starting condition, we say it is highly metamorphosed; if it has changed only a little, we say it is slightly metamorphosed. Metamorphism usually occurs due to deep burial of a rock within the earth.

Metamorphic Rock Types Found in Maryland

Slate - very fine-grained, slightly metamorphosed; breaks into even layers.

Phyllite - fine-grained, more metamorphosed than slate, often with a satinlike shine on its surfaces.

Schist - visible mineral grains, moderately metamorphosed; breaks into uneven layers. Schist contains much mica, making it shiny.

Gneiss - visible grains, highly metamorphosed and often color banded with layers of different minerals; does not break into layers well because it contains less mica than schist.

Quartzite - very hard, medium- to coarse-grained, mostly made of quartz; forms by metamorphism of sandstone.

Marble - medium- or coarse-grained, mostly made of calcite; forms by metamorphism of limestone.

Serpentinite - green or black, sometimes fibrous, made of a variety of altered mafic minerals (these minerals as a group are called serpentine). Rocks such as this that are all or nearly all mafic minerals are called *ultramafics*, which may be metamorphic or igneous in origin.

Soapstone - soft, may be layered or unlayered, mostly made of the mineral talc.

Note: If a rock type is slightly metamorphosed, but not enough to really change it into a metamorphic rock, then the prefix "meta-" is applied to the original name. Thus, in Maryland, we find metabasalt, metarhyolite, metagraywacke, metaconglomerate, and other rocks with similar names.

Metamorphic rocks are sometimes described by adding modifiers to the front of the name to indicate the abundance of various minerals; the minerals are listed in order of increasing abundance. For example, in a biotite-quartz-plagioclase gneiss, plagioclase is the most abundant mineral.

Metamorphic and igneous rocks are sometimes referred to together as *crystalline* rocks.

Comparative Resistances for Major Rock Types in Maryland

ROCK TYPE	EFFECT OF MINERALS	EFFECT OF TEXTURE	OVERALL RESISTANCE
granite	moderately resistant	resistant	rather resistant
gabbro	nonresistant	moderately resistant	moderately resistant
pegmatite	resistant	resistant	may be more resistant than granite
conglomerate breccia sandstone	usually resistant	depends on type of cement holding grains together; usually resistant	usually resistant
graywacke	moderately resistant	moderately resistant	moderately resistant
siltstone shale	clay doesn't change but does fall apart	nonresistant, & layers separate easily	nonresistant
limestone dolomite	nonresistant	moderately resistant	moderately resistant to nonresistant
chert	resistant	resistant, but rarely in continuous layers	moderately resistant
slate	moderately resistant	a hard rock, but breaks easily into layers	moderately resistant
phyllite	moderately resistant	softer than slate, less resistant	moderately or less resistant
schist	moderately resistant	breaks into layers	moderately resistant

Appendix C— *continued*

ROCK TYPE	EFFECT OF MINERALS	EFFECT OF TEXTURE	OVERALL RESISTANCE
gneiss	rather resistant	moderately resistant	moderately to highly resistant
marble	nonresistant	moderately resistant	less resistant compared to other metamorphic rocks, but more resistant than limestone
quartzite	very resistant	very resistant	very resistant
metabasalt*	more resistant than basalt due to mineral changes	moderately resistant	moderately resistant

*Other "meta-" rocks are too rare for us to be concerned with here, but most are slightly more resistant than the same unmetamorphosed rock.

Geologic Time Scale

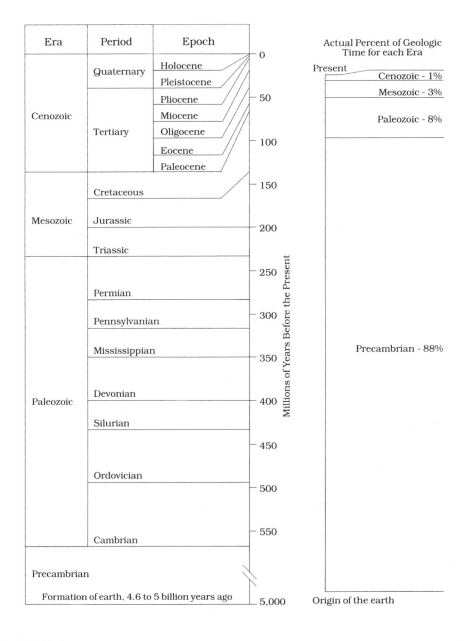

Era	Period	Epoch		Actual Percent of Geologic Time for each Era
Cenozoic	Quaternary	Holocene	0	Present
		Pleistocene		Cenozoic - 1%
	Tertiary	Pliocene	50	Mesozoic - 3%
		Miocene		Paleozoic - 8%
		Oligocene	100	
		Eocene		
		Paleocene		
Mesozoic	Cretaceous		150	
	Jurassic		200	
	Triassic			
Paleozoic	Permian		250	
	Pennsylvanian		300	
	Mississippian		350	
	Devonian		400	
	Silurian			
	Ordovician		450	Precambrian - 88%
			500	
	Cambrian		550	
Precambrian				
Formation of earth, 4.6 to 5 billion years ago			5,000	Origin of the earth

Millions of Years Before the Present

Geologic Column and Summarized History for Maryland

The tables in this appendix summarize the geologic history of Maryland. Each table applies to a different time period, as the titles indicate, with some overlap in time in tables E–1 and E–2. Table E–1 consists of a single chart, continued across pages 142 and 143. Table E–2 begins on page 144 and continues on page 145. Table E–3 is similar, with its first part on page 146 and the remainder on page 147. Thus, in all tables, the oldest time is at the bottom and becomes more recent as you read up a column.

At the left in each chart are the geologic periods or epochs (also found in Appendix D). The "Provinces" columns of tables E–1 and E–2, and the "Coastal Plain Location" columns of table E–3, list the geologic formations of the state in the order in which they formed as one reads up a column. When more than one formation was forming at the same time, there are multiple columns, with more western formations to the left and more eastern ones to the right. The "Environments" column summarizes the depositional environments that existed to create these formations, and the "Plate Movements" column briefly indicates how the movements and conditions of the plates caused the environments. The vertical lines in some areas indicate formations which formed, or events that occurred, over a longer time range.

Since some of the spacing of these tables is determined simply by printing requirements, the tables can give only approximate ages for formations and events. Also, no information on the relative thickness of a formation should be inferred from these tables; that is, several formations which take up one line each are not necessarily the same thickness, and formations in Table E–1 which take up several lines are not necessarily thicker than formations listed on one line. As indicated in the text of Chapter 5, any geologic history is the result of interpretation, and so some parts of these tables may change as new information is collected.

Table E-1—Precambrian to Ordovician Geologic Column and
History Summary

TIME	PROVINCES			
	VALLEY & RIDGE	BLUE RIDGE	WESTERN	PIEDMONT
Ordovician	St. Paul Group Beekmantown Group		Grove	
Early Ordovician to late Cambrian	Conococheague		Frederick	Silver Run Member
Cambrian	Elbrook Waynesboro Tomstown			Marburg Cash Smith
Early Cambrian to late Precambrian	Antietam Harpers	(Antietam - now eroded) (Harpers - now eroded) Weverton	Antietam Harpers	Araby Ijamsville Urbana Sugarloaf
Precambrian		Loudoun Catoctin Swift Run Middletown Gneiss		

PROVINCES			ENVIRONMENTS	PLATE MOVEMENTS
PIEDMONT		EASTERN		
Peach Bottom Cardiff	Ellicott City, Woodstock, & Guilford	Liberty Mélange: Sykesville		Continental blocks pushed together in Taconic orogeny.
	? Glenarm Group:	Morgan Run Conowingo James Run-Port Deposit	Carbonate bank forms along eastern edge of North America.	Subduction below continental fragment produces island arc with related sediments & intrusions.
Gillis — Pleasant Grove Prettyboy	Oella Loch Raven ?	Baltimore Complex Aberdeen		
	Cockeysville		Sediments accumulate on subsided edge & pieces of continent.	
	Setters			Rifting opens Iapetus Ocean; some pieces of continent break off & subside.
Sams Creek			Period of erosion after volcanics. Volcanic activity; Maryland at newly formed eastern edge of North America. Long period of erosion exposes gneisses at surface.	
	Baltimore Gneiss		Metamorphism in Grenville orogeny. Sediments deposited near edge of continent.	Continental collision?

Table E-2—Ordovician to Jurassic Geologic Column and
History Summary

TIME	PROVINCES		ENVIRONMENTS	PLATE MOVEMENTS
	VALLEY & RIDGE, ALLEGHENY PLATEAU	PIEDMONT		
Jurassic	Some of the Newark Group may be early Jurassic, but no other Jurassic rocks appear on the surface in Maryland.		Erosion of Appalachians.	
Triassic		Diabase dikes Newark Group: Gettysburg New Oxford	Intrusions in rift faults. Sediments accumulate in rift basins.	Rifting of continents begins to open Atlantic Ocean.
Permian	Dunkard Group		Deposits in basins on land. All of Maryland from Piedmont west is above sea level from here to present time.	End Alleghany orogeny. Major part of Alleghany orogeny: Final collision of Africa & North America; Western Maryland folded; most rocks in state thrust to the west.
Pennsylvanian	Monongahela Group contains: Waynesburg Coal Pittsburgh Coal Conemaugh Group Allegheny Group contains: Freeport Coal Brookville Coal Pottsville Group		Western Maryland near sea level; frequent alternations between shallow sea, swamp, & land, all collecting sediments.	Early part of orogeny makes highlands in east, sediments flow west; little folding in west yet.
Mississippian	Mauch Chunk Greenbrier Purslane Rockwell			Start Alleghany orogeny.

Table E-2 (cont'd.)

| TIME | PROVINCES | | ENVIRONMENTS | PLATE MOVEMENTS |
	VALLEY & RIDGE, ALLEGHENY PLATEAU	PIEDMONT		
Devonian	Hampshire Foreknobs Scherr Brallier Harrell Mahantango Marcellus Needmore Oriskany Shriver Helderberg		Nonmarine; basin filled by sediments eroded from mountains. Clastics flowing to west. New deep basin forms as continent edge pushes up again. ⊤	End Acadian orogeny. Acadian orogeny: Pressure from NE & SE uplifts eastern Maryland. Start Acadian orogeny.
Silurian	Keyser Tonoloway Wills Creek Bloomsburg McKenzie Rochester Keefer Rose Hill Tuscarora		Sediments accumulate in shallow sea west of eroding Taconic highlands.	End Taconic orogeny.

Ordovician	Juniata
	
	Martinsburg
	
	Chambersburg
	St. Paul Group
	Beekmantown Group:
	  Pinesburg Station
	  Rockdale Run
	  Stonehenge
		Various formations in Piedmont during this time; see Table E-1.	Alluvial sediments from highlands to east fill basin.
			Water deepens as basin forms behind folding continental edge.
			⊤
			Carbonate bank on eastern edge of North America.
			⊥
				Taconic orogeny: Collision of continental fragments & island arc with central North America.
				
				Start Taconic orogeny.

Table E-3—Cretaceous to Present Geologic Column and
History Summary

| PERIOD | EPOCH | COASTAL PLAIN LOCATION | | ENVIRONMENTS | PLATE MOVEMENTS |
		WESTERN SHORE	EASTERN SHORE		
Quaternary	Holocene	Beach deposits	Beach & lagoon deposits	Changes in sea level cause alternating bay or river deposition and erosion, including cutting of river valleys now flooded to create the Chesapeake Bay and its estuaries.	Cretaceous to present: Appalachians may have been slightly uplifted one or more times, thus varying erosion rates and relief.
Quaternary	Pleistocene	Talbot Kent Island Maryland Point Omar Chickamuxen Church	Columbia Group: Parsonburg Sinepuxent Kent Island Ironshire Omar		
Tertiary	Pliocene	Park Hall Upland Gravel 4 Upland Gravel 3 Yorktown Pensauken Brandywine	Walston Silt Beaverdam Yorktown- Cohansey		
Tertiary	Miocene	Eastover St. Marys Choptank Calvert	Chesapeake Group: Manokin St. Marys Choptank Calvert	Marine deposits.	
Tertiary	Oligocene	none in Maryland		No deposits due to low sea levels.	
Tertiary	Eocene	Pamunkey Group: Piney Point Nanjemoy Marlboro		Continental shelf deposits in changing water depths.	
Tertiary	Paleocene	Aquia Hornerstown Brightseat			

PERIOD	EPOCH	COASTAL PLAIN LOCATION WESTERN SHORE EASTERN SHORE		ENVIRONMENTS	PLATE MOVEMENTS
Cretaceous		Monmouth Matawan Group		Changing water depth.	
		Magothy Potomac Group: Patapsco Arundel Patuxent		Shoreline sediments. Nonmarine floodplain.	Atlantic Ocean opening; began building of Coastal Plain & Continental Shelf.

Rock and Sediment Types of Named Formations in Maryland

Allegheny Plateau, Valley and Ridge, and Coastal Plain Formations

Quaternary

Talbot - clay-silt and sand.

Columbia Group

Parsonburg Sand - light-colored sand and some peat.

Sinepuxent - silty sand, with some clay and peat.

Kent Island - interbedded silt, clay, and sand.

Ironshire/Maryland Point - sand, and silty to clayey sand.

Omar - clay and silt.

Chickamuxen Church - gravel and sand.

Tertiary

Walston Silt/Park Hall - silty and clayey sand, with some gravel.

Beaverdam/Upland Gravel 4 - gravel and sand, sometimes muddy or silty.

Upland Gravel 3 - gravel and sand.

Yorktown/Yorktown-Cohansey - clayey silt and sands.

Pensauken - sand and gravelly sand with some silt and clay.

Brandywine - sand, pebbly sand, gravel, and some clay.

Chesapeake Group

Manokin/Eastover - sand and gravel aquifers, with clay or silt impermeable beds.

St. Marys - sandy clay; abundant marine fossils.

Choptank - yellowish sand, clayey silt, and abundant marine fossils.

Calvert - sand, silt, clay; abundant marine fossils.

Pamunkey Group

    Piney Point - sand, with shell layers.

    Nanjemoy - sand, silt, beds or lenses of silty clay; some fossils.

    Marlboro Clay - clay with silt.

    Aquia - sand with some clay, sand cemented by calcite; marine
        fossils.

Hornerstown - mostly glauconite, a green, fine-grained mica.

Brightseat - sand with some clay; marine fossils.

Cretaceous

Monmouth - sand, sometimes iron-cemented; marine fossils.

Matawan Group - sand, with clay and silt; marine fossils

Magothy - sand, with layers of silty clay and gravel; fossil leaves.

Potomac Group

    Patapsco - sand, silt, clay, gravel; fossil plants.

    Arundel - clay, with iron-cemented sandstone; fossil leaves and
        dinosaur bones.

    Patuxent - sand, gravel, some clay; fossil leaves.

Triassic

Newark Group

    Gettysburg Shale - red shale, some sandstone and mudstone;
        fossil leaves and dinosaur footprints.

    New Oxford - red sandstone, some siltstone, shale, and conglom-
        erate; a few reptile bones and teeth.

Permian

Dunkard Group - shale, siltstone, sandstone, some impure coal and
    fresh-water limestone, conglomerate at base.

Pennsylvanian

Monongahela Group - shale, sandstone, limestone, coal.

Conemaugh Group - shale, sandstone, limestone, coal; marine
    fossils.

Allegheny Group - shale, sandstone, siltstone, coal, conglomerates;
    plant fossils.

Pottsville Group - sandstone and conglomerates, with thin siltstone
    and shale; plant fossils.

Mississippian

Mauch Chunk - red and green sandstone; siltstone and shale.

Greenbrier - muddy limestones, limey sandstones and shales; abundant fossils, especially brachiopods.

Purslane - sandstone and conglomerate; thin shaley coal and red shale.

Rockwell - sandstone, conglomerate, shales, siltstones, and coal.

Devonian

Hampshire - red and green sandstones and shales; some siltstones and mudstones.

Foreknobs - sandstone and conglomerate; some shales; abundant marine fossils.

Scherr - conglomeritic sandstones and sandy shales; abundant fossils in some parts of the formation.

Brallier - shale and siltstones; marine fossils.

Harrell Shale - black shales; some imprint fossils.

Mahantango - shale, siltstone, some sandstone; abundant fossils, especially brachiopods, pelecypods, and gastropods.

Marcellus - black shale; fossils on some bedding planes.

Needmore - dark shale and clayey limestone; imprint fossils.

Oriskany Sandstone - pure quartz sandstone, with some lime cement; abundant fossils, especially brachiopods and gastropods.

Shriver Chert - dark shale and dark fine-grained quartz.

Helderberg Group - limestones, with some shale; abundant fossils.

Silurian

Keyser Limestone - limestone, some limey shale; fossils, especially brachiopods and corals.

Tonoloway Limestone - limestone, dolomite, and limey shale; abundant fossils.

Wills Creek - limey shales, mudstone, clayey limestone; mud cracks and ostracod fossils.

Bloomsburg - red sandstone and shale; clayey limestone.

McKenzie - shales and muddy limestone; abundant diverse fossils.

Rochester Shale - limestone and limey shales.

Keefer Sandstone - sandstone, more limey to the west.

Rose Hill - shale and iron-cemented sandstone; varied fossils.

Tuscarora Sandstone - massive silica-cemented sandstone; trace fossils only.

Ordovician

Juniata - red and green sandstone and conglomerates, with siltstone, shale, and graywacke.

Martinsburg Shale - dark shale, with graywackes and siltstones; diverse fossils on some bedding planes.

Chambersburg Limestone - muddy limestone; abundant diverse fossils.

St. Paul Group - limestones; mollusc and brachiopod fossils.

Beekmantown Group

Pinesburg Station Dolomite - gray dolomite.

Rockdale Run - dolomite, and muddy limestone; algae fossils.

Stonehenge Limestone - muddy limestone; abundant fossils.

Cambrian

Conococheague Limestone - shaly limestone; some conglomerate, sandstone, and dolomite; algae, trilobite, and brachiopod fossils.

Elbrook Limestone - shaly limestone and limey shales.

Waynesboro - sandstone, shale, dolomite.

Tomstown Dolomite - dolomite and limestone; a few fossils.

Antietam - pure quartz sandstone; trilobite and brachiopod fossils.

Harpers - rocks of differing metamorphism: shale, slate, phyllite.

Blue Ridge and Piedmont Formations

Formations in this section are grouped by their columns in Table E-1; for ages, see Table E-1.

Blue Ridge

Weverton - quartzite.

Loudoun - conglomerate, phyllite, slate.

Catoctin - metamorphosed basalt, tuff, and rhyolite.

Swift Run - quartzite, slaty metatuff, and marble; some metarhyolite and breccias.

Middletown Gneiss - granitic gneiss, with metadiabase dikes.

Western Piedmont

Grove Limestone - limestone, and some dolomite.

Frederick Limestone - thin-bedded muddy limestone.

Marburg - phyllite and quartzite. Silver Run Limestone member - thin-bedded schistose limestone.

Cash Smith - phyllite, slate, metashale.

Araby - siltstone and silty shale, slightly metamorphosed by shearing forces.

Ijamsville - phyllites and some quartzites.

Urbana - phyllites and thin quartzites.

Sugarloaf - white quartzite.

Gillis - phyllite.

Sams Creek - metabasalt, phyllite, and marble.

Eastern Piedmont

Peach Bottom Slate - blue-black slate.

Cardiff Metaconglomerate - micaceous quartz-pebble metaconglomerate.

Pleasant Grove - schist and some quartzite.

Prettyboy - schist and metagraywacke.

Ellicott City, Guilford, and Woodstock - granites, each varying somewhat in relative amounts of mafics and felsic minerals.

Glenarm Group

Oella and Loch Raven - biotite-plagioclase-muscovite-quartz schists, with garnet and other minerals.

Cockeysville - marble, metadolomite, and limey schists.

Setters - quartzite with mica on bedding planes; some mica schist.

Baltimore Gneiss - gneisses, variable in composition from granitic to quite mafic; variable in thickness and amount of layering.

Liberty Mélange - a subduction zone mixture in general.

Sykesville - gneiss and schist, mafic breccias, metagraywacke.

Morgan Run - schist and metagraywacke.

Conowingo Diamictite - gneiss, breccias, and metamorphosed mixture of sediments.

James Run-Port Deposit - granitic gneisses; metabasalt and other metavolcanic rocks.

Baltimore Complex - gabbro and metagabbro, and ultramafic rocks such as serpentinite.

Aberdeen Gabbro - metagabbro and other mafic rocks.

Appendix G

References and Sources of Geological Information

Selected References

American Geological Institute, *Dictionary of Geological Terms,* New York, NY: Anchor Books, 1976.

Bonin, William A., *The Mineral Industry of Maryland in 1985,* Baltimore: Maryland Geological Survey, Information Circular No. 44, 1986.

Cardwell, Dudley H., *Geologic History of West Virgina,* West Virginia Geological and Economic Survey, 1975.

Cleaves, Emery T., *Physiographic Map of the White Marsh Quadrangle,* Baltimore: Maryland Geological Survey, 1979.

Cleaves, Emery T., Jonathan Edwards, Jr., and John D. Glaser, *Geologic Map of Maryland,* Baltimore: Maryland Geological Survey, 1968.

Chew, V. Collins, *Underfoot: A Geologic Guide to the Appalachian Trail,* Harpers Ferry, WV: Appalachian Trail Conference, 1988.

Dott, Robert H., Jr., and Roger L. Batten, *Evolution of the Earth,* New York: McGraw-Hill, 1981.

Eicher, Don L., *Geologic Time,* Englewood Cliffs, NJ: Prentice-Hall, 1976.

Edwards, Jonathan, Jr., *Geologic Map of the Union Bridge Quadrangle,* Baltimore: Maryland Geological Survey, 1986.

Friedman, Gerald M., and John E. Sanders, *Principles of Sedimentology,* New York: John Wiley & Sons, 1978.

Frye, Keith, *Roadside Geology of Virginia,* Missoula, MT: Mountain Press Publishing Co., 1986.

Gates, A.E., P.D. Muller, and D.W. Valentino, "Terranes and Tectonics of the Maryland and Southeast Pennsylvania Piedmont," in Schultz, A., and E. Compton-Gooding, eds., *Geologic Evolution of the Eastern United States,* Martinsville, VA: Virginia Museum of Natural History, Guidebook No. 2, 1991.

Glaser, John D., *Geology and Mineral Resources of Southern Maryland*, Baltimore: Maryland Geological Survey, Report of Investigations No. 15, 1971.

Hatcher, R.D., Jr., W.A. Thomas, and G.W. Viele, eds., *The Appalachian-Ouachita Orogen in the United States*, Boulder, CO: Geological Society of America, Vol. F-2, *The Geology of North America*, 1989.

Higgins, Michael W., and Louis B. Conant, *The Geology of Cecil County, Maryland*, Baltimore: Maryland Geological Survey, Bulletin 37, 1990.

Hunt, Charles B., *Natural Regions of the United States and Canada*, San Francisco: W.H. Freeman & Co., 1974.

McLennan, Jeanne D., *Dinosaurs in Maryland*, Baltimore: Maryland Geological Survey, 1973.

Mid-Atlantic Region Geological Highway Map, Tulsa, OK: American Association of Petroleum Geologists, 1989.

Muller, Peter D., and David A. Chapin, "Tectonic Evolution of the Baltimore Gneiss Anticlines, Maryland," in *The Grenville Event in the Appalachians and Related Topics*, Boulder, CO: Geological Society of America, Special Paper 194, 1984.

Petersen, Morris S., et al., *Historical Geology of North America*, Dubuque, IA: Wm. C. Brown Co., 1980.

Prosser, L.J., Jr., *The Mineral Industry of Maryland in 1987*, Baltimore: Maryland Geological Survey, Information Circular No. 47, 1987.

Rankin, Douglas W., "The Continental Margin of Eastern North America in the Southern Appalachians: The Opening and Closing of the Proto-Atlantic Ocean," *American Journal of Science* 275-A (1975): 298-336.

Shreve, Forrest, et al., *The Plant Life of Maryland*, Baltimore: The Johns Hopkins University Press, 1910.

Stanley, Steven M., *Earth and Life Through Time*, New York: W.H. Freeman & Co., 1986.

Stearn, Colin W., et al., *Geological Evolution of North America*, New York: John Wiley & Sons, 1979.

Van Diver, Bradford B., *Roadside Geology of Pennsylvania*, Missoula, MT: Mountain Press Publishing Co., 1990.

Volkes, Harold E., and Jonathan Edwards, Jr., *Geography and Geology of Maryland*, Bulletin 19, Baltimore: Maryland Geological Survey, 1974.

Walker, Patrick N., *Water in Maryland: A Review of the Free State's Liquid Assets*, Baltimore: Maryland Geological Survey, 1970.

Sources of More Information on Geology

Other publications on Maryland geology, at varying levels of detail, are available from: Maryland Geological Survey, 2300 St. Paul Street, Baltimore, MD 21218. Phone: 410-554-5505

For geology of the United States: *Natural Regions of the United States & Canada,* by Charles B. Hunt, published by W. H. Freeman & Co.

Publications on all subjects relating to United States geology, and maps, are available from: Earth Science Information Center, U.S. Geological Survey, 507 National Center, Reston, VA 22092. Phone: 703-648-6892 or 703-648-6045

Contacts with Maryland clubs or associations relating to geology can be found through the Maryland Science Center Education Department, phone: 410-685-2370.

Poster-size color versions similar to Fig. 1-1 are available from: Earth Satellite Corporation, 6011 Executive Boulevard, Suite 400, Rockville, MD 20850. Phone: 301-231-0660

Magazines relating to geology:

Earth, Kalmbach Publishing Co., 21027 Crossroads Circle, P.O. Box 1612, Waukesha, WI 53187
Earth in Space, published by the American Geophysical Union, 2000 Florida Ave., NW, Washington, DC 20009
Geotimes, published by the American Geological Institute, 4220 King St., Alexandria, VA 22302
Rocks and Minerals, 4000 Albemarle St., NW, Washington, DC 20016

Geologic maps showing large regions of the United States, in a series of 12 geological highway maps, are available from the American Association of Petroleum Geologists, P.O. Box 979, Tulsa, OK 74101

Index

Page numbers shown in italics indicate the location of definitions for terms used in the text.